Burkhard Freiherr von Wangenheim

Survival trees – a new method in innovation theory

A successful introduction of a method commonly used in survival analysis into the field of innovation diffusion theory

Anchor Academic Publishing

Freiherr von Wangenheim, Burkhard: Survival trees - a new method in innovation theory: A successful introduction of a method commonly used in survival analysis into the field of innovation diffusion theory. Hamburg, Diplomica Verlag GmbH 2012

ISBN: 978-3-95489-004-0
Print: Anchor Academic Publishing, an Imprint of Diplomica® Verlag GmbH, Hamburg, 2012

Bibliographical Information of the German National Library:
The German National Library lists this publication in the German National Bibliography. Detailed bibliographic data can be found at: http://dnb.d-nb.de

The digital publication (eBook) of this work with the ISBN 978-3-95489-504-5 can be purchased on the general market or directly from the publisher.

All rights reserved. This publication may not be reproduced, stored in a retrieval system or transmitted, in any form or by any means, electronic, mechanical, photocopying, recording or otherwise, without the prior permission of the publishers.

Dieses Werk ist urheberrechtlich geschützt. Die dadurch begründeten Rechte, insbesondere die der Übersetzung, des Nachdrucks, des Vortrags, der Entnahme von Abbildungen und Tabellen, der Funksendung, der Mikroverfilmung oder der Vervielfältigung auf anderen Wegen und der Speicherung in Datenverarbeitungsanlagen, bleiben, auch bei nur auszugsweiser Verwertung, vorbehalten. Eine Vervielfältigung dieses Werkes oder von Teilen dieses Werkes ist auch im Einzelfall nur in den Grenzen der gesetzlichen Bestimmungen des Urheberrechtsgesetzes der Bundesrepublik Deutschland in der jeweils geltenden Fassung zulässig. Sie ist grundsätzlich vergütungspflichtig. Zuwiderhandlungen unterliegen den Strafbestimmungen des Urheberrechtes.

Die Wiedergabe von Gebrauchsnamen, Handelsnamen, Warenbezeichnungen usw. in diesem Werk berechtigt auch ohne besondere Kennzeichnung nicht zu der Annahme, dass solche Namen im Sinne der Warenzeichen- und Markenschutz-Gesetzgebung als frei zu betrachten wären und daher von jedermann benutzt werden dürften.

Die Informationen in diesem Werk wurden mit Sorgfalt erarbeitet. Dennoch können Fehler nicht vollständig ausgeschlossen werden, und der Diplomica Verlag, die Autoren oder Übersetzer übernehmen keine juristische Verantwortung oder irgendeine Haftung für evtl. verbliebene fehlerhafte Angaben und deren Folgen.

© Anchor Academic Publishing ein Imprint der Diplomica® Verlag GmbH
http://www.diplomica-verlag.de, Hamburg 2012
Printed in Germany

Table of contents

1 - Introduction 7
  1.1   Context of Thesis 7
  1.2   Contribution of Thesis 10
  1.3   Structure & Internal Pattern of Thesis 11

2 - Modelling Censored Event Data in the Context of Innovation Adoption- and Diffusion Theory 13
  2.1   Analysing and Forecasting Innovation Diffusion by Dynamic Micro Models 14
  2.2   General Concepts and Terminologies 19
  2.3   Statistical Framework 23
  2.4   Classical Methods for the Analysis of Event History Data 29
      2.4.1   Non-Parametric Methods 29
      2.4.2   Parametric Methods 35
      2.4.3   Semi- Parametric Methods 37

3 - Presentation and Analysis of the Survival Tree Method 40
  3.1   Review of CART^{TM} 41
  3.2   Principle Framework & Mechanics of Survival Trees 46
  3.3   Splitting, Pruning, Tree-Selection & Alternative Proposals for Survival Trees 47
      3.3.1   Splitting – Growing the Saturated Tree 48
      3.3.2   Pruning – Generation of Optimal Subtree Sequence 56
      3.3.3   Final Tree Selection 58
      3.3.4   Alternative Approaches 59
  3.4   Final Assessment 60
      3.4.1   Assessment of the Splitting, Pruning and Tree Selection Proposals 60
      3.4.2   Merits & Deficiencies of the Survival Tree Method 63

4 - The use of Survival Trees to Forecast Innovation Diffusion 68
  4.1   Applicability of Available Software 69
  4.2   Data Description & Handling 71
  4.3   Implementation 74
  4.4   Results 75
  4.5   Discussion 81

5 - Summary 87

6 - Appendix **88**
6.1   Classification and Regression Tree for E-purchase Adoption 88
6.2   Cross Table for Sector and Country Coverage 88
6.3   Variable Description and Handling.. 89
6.4   R-Syntax for Survival Tree 90
6.5   Data Output for E-purchase Survival Tree 91
6.6   Saturated Survival Tree for E-purchase Adoption 92
6.7   Original Survival Tree for E-purchase 92

7 - Indices **93**
7.1   Index of Abbreviations 93
7.2   Index of Symbols 93
7.3   Index of Synonyms 94
7.4   Index of Tables 94
7.5   Index of Figures 94

8 - Bibliography **96**

1 Introduction

1.1 Context of Thesis

"It is almost universally accepted that technological change and other kinds of innovations are the most important sources of productivity and increased material welfare – and that this has been so for centuries".[1]

On the corporate level, the recognition has succeeded that the implementation and maintenance of a successful innovation management is the key contribution to competitiveness and future growth. For this reason, there is great interest in understanding the processes of innovation and its subsequent diffusion to formulate appropriate policies.

Within the last decades, researchers in management and marketing science have greatly contributed to the development adoption- and diffusion theory by suggesting analytical models for describing and forecasting the diffusion of an innovation in a social system. The main reason for this has been the perceived high failure rate of new products and the consequent needs to improve the related management and marketing decisions.

The explanation why firms do not instantaneously adopt new technology immediately after its commercialisation (i.e. diffusion is a time-intensive process) can be traced to different theories of innovation diffusion advocated in literature. According to early epidemic theories of inter-firm diffusion,[2] diffusion is a disequilibrium process resulting from information symmetries between potential adopters.[3] In contrast to epidemic models, contemporary approaches to technology diffusion are characterised by the dismissal of information spreading as the key explanatory variable of innovation diffusion.[4] Rather, models in general assume that firms behave optimally (i.e. are profit maximizers) and that information pertaining to the technological and economic characteristics of the information is perfect. Within this equilibrium approach there are three categories of models that have been developed in the literature: the rank or

---

[1] Charles Enquist (1997).
[2] The economy wide-degree of diffusion can be decomposed into two elements: Inter-firm diffusion and intra-firm diffusion. Inter-firm diffusion describes a firm's first use of a new technology. Intra-firm diffusion, on the other hand, has not been researched much so far and describes the increasing intensity of technology diffusion. See for literature on inter- and intra diffusion Griliches (1957), Mansfield (1968), Bass (1969) and Hollenstein, Wörter (2004), respectively.
[3] Baptista (2000).
[4] Gourlat, Pentecost (2000).

probit; stock or game theoretic, and order effects models.

In rank or probit models[5] potential adopters of technology have different inherent characteristics and as a result obtain different gross returns from its use.

The essence of stock effect models is that benefits to the marginal adopter from acquisition decreases as the number of previous adopter's increases.[6]

Order effect models are similar to the rank effect models in that the gross returns of a firm adopting a new technology depends upon its position in the order of adoption, with higher order adopting firms achieving a greater return than low-adopters.[7]

Despite the continuing progress of contemporary approaches, the main impetus underlying diffusion research is still the epidemic Bass model[8]. Subsuming the majority of other models derived from that model or independently, this model addresses the market in the aggregate. The typical variable measured is the number of adopters who purchase the product by a certain time t. The emphasis is on the total market response rather than on the individual adopter. Here, the individual characteristics of potential adopters and their impact on the decision-process remain wholly uncovered. Not the individual who decides, whether to adopt or reject an innovation is central to the analysis, but the time-related distribution of the adoption decision dependent on marketing variables.[9] These models cannot explain why a particular individual adopts or rejects an innovation at a specific point in time. Consequently, these models achieve no adequate aggregation of individual adoption decisions. Although the specific managerial implications that these models give should not be questioned in general, they remain limited by the aggregate perspective which they take.

In fact, diffusion theory faces a constant dilemma between disaggregate and aggregate diffusion modelling. Although it is unquestionable that the diffusion process is built upon individual adoption decisions, the persuasion that diffusion models should thus be built upon individual decisions has not yet fully materialized. One reason lies undoubtedly in the substantial modelling obstacles that theory has faced so far in trying to pursue this.

---

[5] Ireland, Stoneman (1986).
[6] Reingannum (1981a, 1981b, 1989), Quirmbach (1986).
[7] Gourlay, Pentecost (2000), p. 3.
[8] Bass (1969).
[9] Albers (1998), p. 13, Kühnapfel (1995), p. 121.

Most models that allow for illumination of individual adoption behaviour are static in nature, hereby failing to capture the inherent dynamics of the diffusion process which makes plausible aggregation nearly impossible. *This dilemma has forced an explicit distinction between adoption- and diffusion theory. Although this distinction is often taken to frame the sort of analysis that is performed, it is forced by the disability of most diffusion models to persuasively incorporate the naturally inherent individual perspective.*

By recognizing that the diffusion process is built upon individual adoption decisions, the adoption theory should be recognized and modelled much more as the key basement of diffusion theory rather than a theory that is conceptionally and in content different to the diffusion theory. The implication of this is that diffusion models that take the individual perspective simultaneously perform an adoption analysis.

Moreover, diffusion models based on individual adoption decisions offer an opportunity to study the actual pattern of social communication and its impact on product perceptions, preferences and ultimate adoption. Nonetheless, first attempts to establish the diffusion process on the basis of individual adoption decisions faced severe problems in realizing ultimate aggregation.[10] Merely the study by Chatterjee and Eliashberg (1989) provided encouraging empirical evidence for a useful aggregation of individual adoption decisions.[11] Indeed, it has been recognized only recently that the above described dilemma can be solved.

So-called event history data is able to capture the dynamics of the diffusion process while, simultaneously, the individual perspective (micro level) can be preserved.

Eventually, with the introduction of hazard models[12] into diffusion theory, various micro models were found that could effectively deal with event history data and thus allowed for consideration of individual heterogeneity among adopters by incorporating covariate effects into diffusion models. Up to now, most models that have come up in the widely applied field of event history analysis have been applied to diffusion theory, too.[13] It should be said, however, that these applications have taken place only recently making the use of event history data still a novel thought to diffusion theory.

---

[10] Hiebert (1974), Stoneman (1981), Feder, O`Mara (1982), Jensen (1982).
[11] Mahajan, Muller, Bass (1990).
[12] Kalbfleisch, Prentice (1980), Cox, Oakes (1984).
[13] Reingannum (1982), Hannan, Mc Dowell (1984, 1987, 19990), Sinha, Chandrashekaran (1992), Gönül, Srinivasan (1993), Caudil et al. (1995), Gourlat, Pentecost (2000), Litfin (2000).

The main reason for this may lie in the extent of data collection necessary to perform an analysis. Especially, in economic theory, where the necessity for event history data is not obvious, this may prove a vital obstacle; keeping track of each individual and his adoption decision is undoubtedly a more challenging task than simply taking the aggregate approach. Fortunately, with rising technological possibilities, the applicability of event history models has risen, too.

1.2   Contribution of Thesis

With the extension of the non-parametric classification and regression tree method (CART^{TM})[14] to the analysis of censored event data, we are now given the opportunity to move research forward by examining usefulness and applicability of that method for the analysis and forecast of innovation diffusion. The development of the so-called "survival trees" was highly motivated by the need to develop meaningful prognosis rules in medical science.[15] As will be shown later, there are a number of essential parallels between survival analysis in medical science and diffusion analysis in economics. Emergences of new methods in that field are therefore likely to prove applicable in adoption- and diffusion theory (ADT), too.

As the CART^{TM} method itself is still new to economic theory, it should not surprise that no known application of survival trees has taken place in an economic context so far. Indeed, even for the CART^{TM} method only two applications in an economic context are known.[16] Both methods, CART^{TM} and survival trees, have been developed in the area of medical science and seem to spread only slowly to other scientific areas. Economists and other non-medical scientists alike will have to be persuaded of the new insights that these methods offer. As for survival trees, this thesis is the first attempt to do this.

The method offers additional insights into causal relations that traditional methods fail to give and can therefore resemble a powerful contribution to modern diffusion theory. Its interpretational power makes it likely that this method will meet widespread acceptance.

---

[14] Breiman et al. (1984).
[15] Gordon, Olshen (1985).
[16] Haughton, Oulabi (1993), Köllinger, Schade (2004).

1.3 Structure & Internal Pattern of Thesis

I want to briefly put into words the structure of the thesis that is already summarized in the table of contents. I believe this will make it more easily understandable and more coherent. Additionally, I find it important that the reader is aware of the internal pattern underlying this thesis. With this, I mean simple formatting or used terminology decisions.

Let us start with the **structure**: In the course of the thesis, the survival tree method will be introduced within the context of ADT. For this reason, I will provide arguments in favour of dynamic micro models as a means to analyse and forecast innovation diffusion (section 2.1).

As event history data enables us to do this, I will set up the common concepts and ideas of event history data modelling just as the classical methods from this area, all within the context of ADT (2.2, 2.3, and 2.4). This will be done to grasp an understanding of the interpretation and functionality of the event history patterns within the ADT context and is considered essential for understanding the survival tree method and its usefulness in forecasting innovation diffusion.

Survival trees have been derived from CART^{TM} and consequently both methods share essential conceptual features. After a general introduction into the CART^{TM} methodology (3.1) and a first introduction in the area of survival trees (3.2), I will attempt to classify the proposals that have come for the construction of survival trees into three building blocks that are commonly used in the construction of CART^{TM} (3.3).

Subsequently, the various proposals that have come up in the construction of survival trees will be evaluated and the merits just as the deficiencies of the method will be discussed (3.4).

I will describe in detail the software applications available for survival tree calculations to facilitate future work on them (4.1). The data that the method will be applied on is presented and the way data was handled is documented (4.2) before I state which of the various options was taken (4.3).

Analysing the results, we will see whether the method can offer new insights into ADT and whether the previously discussed merits & deficiencies of the method hold true or might have to be reconsidered in the discussed context (4.4).

Eventually, I will discuss the central question about the usefulness of the method to forecast innovation diffusion. I will try to relate the method's results and their implications to economic practice. Other related issues and thoughts will be discussed, as well (4.5). Conclusively, main patterns and findings of the thesis will be summarized (4.6).

Let me now explain the **internal pattern** of the thesis relating to measures that were taken to ease functionality and readability of the thesis.

The problem of inconsistent terminology is particularly apparent in event history analyses. If we take, for instance, the denomination "event history data", we can easily find at least five other denominations, all used interchangeably, which may sometimes hamper understanding substantially. I will thus name these cases when they appear and say explicitly which of the various denominations I will use. Additionally, I have developed an index of synonyms in Appendix 7.3 to prevent any confusion.

Other confusion is likely to be caused by the various denominations in ADT. No definite rule can be established as to whether one should use adoption theory or diffusion theory for a specific field under investigation. In this thesis, I claim that these two areas belong essentially together. I will therefore make no distinction between these two areas using the single denomination adoption- and diffusion theory (ADT) throughout this thesis.

Besides, there is no generally agreed structure in the area as to what model belongs to what class of models and so on. The classification of models into micro and macro, static and dynamic models is by no means generally agreed and was adapted from Litfin (2000).

For easier readability and in order to put emphasize on sentences that I consider vital, I will format respective text **bold** or *italic*. In this way, words representing important issues are formatted **bold** to enable easier localization.[17] Italic formatting is used for sentences *that I considered vital for overall understanding*.

I have noticed that the literature on survival trees has picked up momentum within the year 2003 and 2004, especially. This made it difficult to incorporate all new literature in the thesis as it was published while this thesis was written. Yet, I think I

---

[17] Bold was used for the authors of the various proposals in 3.3. because their names stand exemplarily for the method they developed.

have successfully attempted to include all literature until the end of November 2004 in the thesis.

Sometimes, I will sum up findings or provide a brief outlook at the very beginning of a section. I do this to make sure one does not lose track of the findings and is always aware of why a certain section was written.

2 Modelling Censored Event Data in the Context of Innovation Adoption- and Diffusion Theory

In virtually every area of the social sciences, there is great interest in events and their causes. Criminologists study crimes, arrests, convictions, and incarcerations. Medical sociologists are concerned with hospitalizations, visits to a physician and psychotic episodes.[18] As a field of economics, innovation theory investigates and tries to predict the effects of innovations on society. Hereby, the adoption decisions of the members of society play the decisive role.

In each of the above mentioned examples, an event consists of some qualitative change that occurs at a specific point in time. Because events are defined in terms of change over time, it has become increasingly acknowledged that the best way to study events and their causes is to collect event history data.[19] In its simplest form, event history is a "longitudinal record of when events happen to a sample of individuals or collectivities"[20].

In this chapter, I will provide reasons why innovation diffusion analysis and forecast should be performed on the basis of dynamic micro models. These models can be established only on the basis of event history data. As all models from the area of event history analysis are either directly or indirectly based on the hazard rate framework, I will establish this framework to ease understanding of the upcoming presentation of the various parametric, semi-parametric and non-parametric models.

For the upcoming introduction of the survival trees, *it is important to understand the conceptionel parallels between diffusion theory and survival analysis*. These parallels allow us to use models coming from the area of survival analysis for ADT.

---

[18] Allison (1984) p. 9.
[19] Alternatively, data is collected as cross-sectional or panel data. For a comparison of these approaches with event history data collection see Blossfeld, Rohwer (2002), pp. 4-6.
[20] Allison (1984) p.9.

2.1 Analysing and Forecasting Innovation Diffusion by Dynamic Micro Models

"An innovation is an idea practice or object that is perceived as new by an individual or another unit of adoption"[21]. Commonly speaking, innovation diffusion theory addresses how new ideas, products and social practices spread within society or from one society to another. Moreover, adoption theory analyzes the process of innovation adoption by an individual. Both theories aim to identify explanatory variables that drive and determine the respective process. The adoption process of each individual can differ in starting point and duration. In this way, adoption decisions of members of social systems are spread across time. *Consequently, the adoption theory forms the fundament of innovation diffusion theory and is thus part of it.*

While, by definition, adoption theory is mainly concerned with the exploration of the determinants of adoption, the diffusion theory focuses on the aggregate analysis of all adoption decisions of the members of a social system.

However, by recognizing that the diffusion process is built upon individual adoption decisions, the adoption theory should be recognized and modelled much more as the key basement of diffusion theory rather than a theory that is conceptionally and in content different to the diffusion theory. For this reason, I will make no explicit distinction between these two theories which I claim to belong together.[22]

The diffusion of an innovation has traditionally been defined as the process by which "an innovation is communicated through certain channels over time among the members of a social system"[23]. This definition, with its reference to innovation, communication (and the respective communication channels), time and the members of a social system names the four key components widely recognized as driving innovation diffusion. Although the diffusion process is undoubtedly a dynamic process, the majority of the models that have emerged in diffusion theory could only insufficiently capture this essential feature.[24] Empirical research for analysis and forecast of the diffusion process is still dominated by aggregate diffusion models that mostly envisage capturing the influence of marketing variables on the success of an innovation.

---

[21] Rogers (1995), p. 11.
[22] We will refer to "adoption- diffusion theory" (ADT).
[23] Rogers (1995), p. 5.
[24] Litfin (2000), p. 21.

These approaches are convenient in practical terms but they raise the following question: Can a genuine diffusion model be constructed by aggregating demand from consumers who behave in the neoclassical way? That is, assume that consumers are smart and are not just carriers of information? They therefore maximize some objective function such as expected utility or benefit from the product, taking into account the uncertainty associated with their understanding of its attributes, its price, pressure from other adopters to adopt it and their budget. Because the decision to adopt is individual-specific, all potential adopters do not have the same probability of adopting the innovation in a given time-period. Is it possible to develop the adoption curve at the aggregate market level, given the heterogeneity among potential adopters in terms of adopting the innovation at any time t?[25]

In fact, aggregate models cannot explain why an individual adopts or rejects an innovation at a specific point in time. As a result, these models achieve no adequate aggregation of individual adoption decisions. Analysis and forecast of adoption procedures by means of these models is hardly convincing. While attempts have been taken to unbundle adopters of the aggregate level by categorizing adopters ex-post into a scheme, they could not eliminate the shortcomings of the underlying assumption of adopter homogeneity.[26]

The general scheme used for adopter classification is that of Rogers. Rogers divided individual responses to technology into five ideal categories: innovators, early adopters, early majority, late majority, and laggards.[27] According to him, the main concern of the innovation diffusion research is how innovations are adopted and why innovations are adopted at different rates. Furthermore, he identified five characteristics of innovations that help to explain differences in adoption rates: relative advantage, compatibility, complexity, trialability, and observability. His work has become fundamental to innovation diffusion research and has been documented and quoted in many papers and books.

Although a wide variety of innovations and diffusion processes have been investigated, one research finding keeps recurring. If the cumulative adoption time path or temporal pattern of the diffusion process is plotted, the resulting distribution can generally be described as taking the form of an s-shaped (sigmoid) curve.[28] The

---

[25] Following Mahajan, Muller, Bass (1990), p. 6.
[26] Mahajan, Muller, Bass (1990), p.6.
[27] See Rogers (1983), pp. 244-245 for a detailed description of the 5 adopter categories.
[28] The original diffusion research was done as early as 1903 by the French sociologist Gabriel Tarde who plotted

observed regularity in the diffusion process results from the fact that initially only few members of the social system adopt the innovation in each time period. In subsequent time periods, however, an increasing number of adoptions per period occurs as the diffusion process begins to unfold more fully. Finally, the trajectory of the diffusion curve slows down and begins to level off, ultimately reaching an upper asymptote. At this point diffusion is complete.[29]

In entrepreneurial reality, information about the process of diffusion is crucial to the success of new product marketing. If this information is provided on the aggregate level, however, marketing implications are limited. A company will not know whom to target to drive the diffusion process forward. These shortcomings may have let to an unquantifiable waste of resources as companies are likely to have targeted late adopters in the early stages of the innovations market placing and vice-versa. *A tool that can identify crucial target groups at every stage of the diffusion process is seen to be of utmost importance in marketing. So far, there is no method that is capable of providing this insight.*

Besides, the witnessed unilateral reliance on aggregate models may have let to a great number of incorrect diffusion prognoses. The most prominent example of an (ex-post) off beam forecast that was based on an aggregate model is described in a diffusion study by Berndt and Altobelli (1991)[30]. Other wrong forecasts may prove the insufficient predictive power of these diffusion models.[31]

In practice, companies need information about target clients and the factors that drive their decisions; something aggregate models cannot provide. This growing recognition has materialized in a mounting demand for rapid integration of micro models to identify and analyse the adoption and diffusion process. Next to the widely used macro models, these micro models can contribute decisively to the analysis and forecast of adoption behaviour and the resulting diffusion process.

Even though the adoption behaviour is nothing but the disaggregated form of the diffusion process, the areas of adoption theory and diffusion theory have been largely separated so far. In fact, not all micro models can be used to analyse and forecast innovation diffusion.

---

the original S-shaped diffusion curve.
[29] Although the diffusion pattern of most innovations can be described in terms of a general S-shaped curve, the exact form of each curve, including the scope and the asymptote, may differ.
[30] Berndt, Altobelli (1991) investigated the failure of BTX screens in Germany
[31] Litfin p. 25.

Generally, all micro models consider the heterogeneity of individuals and allow for the integration of co-variables. There is only one type of model, however, that can adequately model censored event data in order to capture the dynamics of the diffusion process. Thus, I claim that only dynamic micro models can be used to forecast innovation diffusion adequately.

To illustrate this, a comparison between a static model and dynamic micro model will be used.[32]

If the focus of analysis is on finding out whether a specific individual adopts or rejects an innovation at a specific point in time and what explanatory variables can be identified, then logistic regression is often employed.[33] This method explains one dependent dichotomous variable through a number of independent variables. Within the framework of ADT, the dichotomous variable can be labelled "adoption of innovation" and "rejection of innovation" always with respect to one specific point in time.[34] Independent variables could be all sorts of individual characteristics. In ADT one often differentiates between product-, adopter- and environment specific independent variables.[35]

For logistic regression the usual restrictive assumptions that are known from linear regressions have to be taken.[36] A violation of these premises can lead to distorted and inefficient estimations for the regression coefficients and eventually to invalid statistical inferences. Here, empirical research is still severely limited by the existence of multicollinearity and autocorrelation between the independent variables. Generally speaking, logistic regression establishes a functional relation between the probability that an event takes place (i.e. an individual adopts the innovation) and a number of predetermined explanatory variables (i.e. independent variables).

In contrast to the linear regression the observable dependent variable, in this case, is not metric, but dichotomous.[37] Logistic regression quantifies and thus identifies the factors driving or preventing individual innovation adoption. Heterogeneity of individuals is respected and uncovered. Nevertheless, the characteristics of the process itself are not considered at all. With the help of logistic regressions only the result of

---

[32] Other approaches are of game-theoretical (Reinganum, 1983) or econometric nature (Eliashberg, 1990), (Jensen, 1982).
[33] Köllinger, Schade (2004).
[34] "Rejection" should not be understood as being definite but rather with respect to one specific point in time (i.e. at point t_1 the innovation may still be rejected but than be adopted at t_2.
[35] Litfin (2000) p. 25.
[36] Menard (1995) p.4 ff.

the adoption process can be revealed. All individuals who have adopted the innovation in between the market placing and the end of the observation period are classified as adopters. Individuals who have not yet or will never adopt the innovation are accordingly classified as non-adopters. There is no differentiation with respect to the adoption's specific point in time and the future possibility of adoption. *Logistic regression ignores time and thus merely gives a snapshot of adoption behaviour and the diffusion process.*

No valid conclusions can be drawn concerning future market potential, for instance. Despite of this, it is out of the question that with the method elementary relations between adoption decisions and its determinants can be established. Nevertheless, in logistic regressions the duration between market placing and adoption is not taken into account. There is no difference between those individuals who adopt the innovation shortly after market placing and those who adopt shortly before the observation period ends.

Yet, it appears only natural that, by average "early adopters" exhibit a higher likelihood of adoption than "late adopters". The negligence of this information reduces accuracy and inferential power of static estimations.[38] Besides, it is the time-related observation of the adoption process, in particular, that enables predictions about future adoption behaviour and thus innovation diffusion. A solution could be the integration of a time-to-adoption independent variable but then one could only consider the individuals who have already adopted the innovation within the observation period. As for the individuals who have not adopted in the period, no time-to-adoption duration can be asserted, as we do not know when and whether they will adopt the innovation after the observation period ends. These observations are "censored". Censored data can simply be ignored and filtered off the analysis, but this leads to distorted estimates, which is why this approach should be abandoned in the presence of censored data. It is here that the so-called event history models come in.

---

[37] For a detailed description see Aldrich, Nelson (1984).
[38] Allison (1995), p.4.

2.2 General Concepts and Terminologies

The general purpose of the analysis of event history data is to explain why certain individuals are at a higher risk of experiencing the event(s) of interest than others.[39] In general, this can be accomplished by using special types of methods which, depending on the field in which they are applied, are called failure-time models, life-time models, survival models, transition rate models, response-time models or hazard models.[40] It should be noted, however, that the origin of event history data modelling lies in the area of medical science.[41] For this reason and for the continuing dominance of survival analysis within the area of event history data modelling, it is not surprising that all of the models that will be introduced shortly have been developed in this area and thus carry respective denominations.

In hazard models the risk of experiencing an event within a short time interval is regressed on a set of covariates.[42] Two special features distinguish hazard rate models from other types of regression models: They make it possible to include censored observations in the analysis and to use time-varying explanatory variables. Censoring is, in fact, a form of partially missing information: On the one hand, it is known that the event did not occur during a given period of time, but on the other hand, the time at which the event occurred is unknown. Time varying covariates may change their value during the observation period. The ability of including covariates that may change their value in the regression makes it a truly dynamic analysis.

In order to understand the nature of event history data and the purpose of event history analysis, it is important to understand the following four elementary concepts: **state, event, duration, and risk period**. These concepts are illustrated below using first an example from the analysis of unemployment histories.[43]

The first step in event history analysis is to define the relevant **states** which can be distinguished. The states are the categories of the dependent variable, the dynamics of which we want to explain. At every particular point in time, each person occupies exactly one state. In the analysis of unemployment histories, four states are generally distinguished: employment, part-time employment, re-training, and unemployment.

---

[39] Andresen, Keiding (2001), p. 4956.
[40] I will predominantly use the denomination "hazard rate models".
[41] Cox (1995), p.4.
[42] In the context of the survival analysis and event history data, the problem of unobserved heterogeneity /also called selectivity or frailty) has received a great deal of attention. I will not discuss this topic, see Andresen, Keiding (2001), pp. 4956-4962 for more details.

The set of possible states is sometimes called the state space.

An **event** is a transition from one state to another, that is, from an origin state to a destination state. In this context, a possible event is "first employment", which can be defined as the transition from the origin state, unemployed, to the destination state, employed. Other possible events are: taking a part-time employment or a job re-training. It is important to note that the states which are distinguished determine the definition of possible events. If only the states employment and unemployment were distinguished, none of the above mentioned events could have been defined. In that case, the only events that could have been defined would be becoming employed or unemployed.

Another important concept is the **risk period**. Clearly, not all persons can experience each of the events under study at every point in time. To be able to experience a particular event, one must occupy the origin state defining the event, that is, one must be at risk of the event concerned. The period that someone is at risk of a particular event, or exposed to a particular event, is called the risk period. For example, someone can only experience to become unemployed when one was employed before. A strongly related concept is the risk set. The risk set at a particular point in time is formed by all subjects who are at risk at experiencing the event open at that point in time.

Using the concepts, event history analysis can de defined "as the analysis of the **duration** of the non-occurrence of an event during the risk period"[44]. This duration is usually labelled by the term episode[45]. When the event of interest is "first employment", the analysis concerns the duration of non-occurrence of a first employment. In practice, as will be shown below, *the dependent variable in event history models is not duration or time itself but a rate*.

Therefore, event history analysis can also be defined as the analysis of rates of occurrence of the event during the risk period. In the first employment example, an event history model concerns a person's employment rate during the period that he or she is in the state of never having been employed.

---

[43] See for an example of this type of analysis: Heckman, Borjas (1980).
[44] Andersen. Keiding (2001), p. 4957.
[45] Also-called spells, waiting time; one should not become confused by the terminology "the duration of an episode". "Episode" should thus be seen as a purely technical term. I will predominantly use simply "duration".

A strong point of hazard models is that one can use **time-varying covariates**. These are covariates that may change their value over time. Examples of interesting time varying covariates are, in the employment history example, an individual's financial status or health status. As a matter of fact, the time variable and interactions between time and time-constant covariates are time-varying covariates as well.

We now do have to **fit the above described concepts into the area of ADT**:

In ADT one generally distinguishes between two states: "adoption" or "non-adoption" of an innovation. The event will be described by the adoption an innovation, which can be defined as the transition from the origin state, non-adoption, to the destination, adoption. This event pattern is called a "single non-repeatable event" where the term single reflects that the origin state, non-adoption, can only be left by one type of event, and the term, non-repeatable, indicates that the event can occur only once. Models that have been developed for this type of event pattern, we will call **single risk models**[46]. The duration measures the time until an individual adopts an innovation. Logically, an individual does not necessarily have to adopt within the observation period or further beyond it. Individuals that do not adopt within the observation period and of which we do not when or whether at all they will adopt in the time after produce **censored data**. I will describe this phenomenon later within the current context.

By and large, this is the sort of event history pattern that is known from the area of survival analysis. In both fields, we observe the duration that lies between some predefined point in time and one single (absorbing) event. In most cases, survival analysis deals with the investigation of the duration between the beginning of treatment or hospitalization and the death of an individual. Ironically enough, both the adoption decision and the death of an individual are single non-repeatable events.

As both processes are equal in terms of their general event pattern, survival models represent likely alternatives for modelling and analyzing adoption- and diffusion processes. It should thus not surprise that all models that will be introduced come from the area of survival analysis. In effect, the vast number of models in survival analysis has been developed to model this type of event pattern.

There are, indeed, other alternative concepts, some of which may also be used in the context of ADT. I want to shortly introduce these for a more conclusive introduction.

---

[46] This intuitive denomination has been chosen although no previous quotes of it could be found. I believe this will make model distinction much easier.

Sometimes, it may prove necessary or is simply wanted to distinguish between different types of events or risks. In the analysis of death rates, one may, for example, want to distinguish between different causes of deaths. In ADT a distinction between various causes of adoption decisions is equally conceivable.

The standard method for dealing with situations where, as a result of the fact that there is more than one possible destination state, individuals may experience different types of events is the use of **multiple risk or competing risk models**.[47]

Most events studied in social sciences are repeatable, and even most event history data contains information on **repeatable events** for each individual. This is in contrast to medical research and to ADT where the event of greatest interest is death or adoption, respectively. Events of repeatable events could be job changes, having children, arrests, or promotions. In an economic context, the investigation of (repeated) product buying decisions may prove interesting. Often events are not only repeatable but also of different types, that is, we have a **multiple state** situation. When people can move through a sequence of states, events cannot only be characterized by their destination states, as in competing risk models, but they may also differ with respect to their origin state and destination states. An example is, once again, an individual's employment history: An individual can move through the states of employment, unemployment, and out of the labour force. In that case six different kinds of transitions can be distinguished which differ with regard to their origin and destination states.

Hazard models for analyzing data on repeatable events and multiple-state data are special cases of the general family of multivariate hazard models. Another application of multivariate hazard models is the analysis of dependent or clustered observations.[48] Examples are the occupational careers of spouses, educational careers of brothers, child mortality of children in the same family. Hazard rate models can be easily generalized to situations in which there are several origin and destination states and in which there may be more than one event per observational unit.[49]

After this general overview to other event history concepts, it is important to stress again that, in the course of this thesis, I will exclusively introduce and apply models for the analysis of single non-repeatable events (single risk models). Moreover, the

---

[47] See Kalbfleisch, Prentice for an extensive overview over these concepts.
[48] Andersen, Keiding (2001), p. 4960.
[49] See Andersen, Keiding (2001), p.4961 for a general description of the generalization to be performed.

integration of time varying explanatory variables will not be considered.

Therefore, I will model the adoption- and diffusion process as having one origin state, non-adoption, and one final non-repeatable event, adoption. Hereby, I will analyse the impact that time-constant explanatory variables have on the dynamics of this process.

2.3 Statistical Framework

Let me now explain the statistical framework of event history analysis that is essential in understanding hazard models regardless of the specific concept chosen.

As such, hazard models have already been introduced into ADT.[50] Moreover, these models have been used, in an economic context, to analyse and forecast business and firm survival (failure).[51] In hazard models, no time-point related snapshot of adoption behaviour is analyzed but a time-related observation is established that considers the process characteristics. For this purpose, one needs to know of each individual not only whether an event has taken place but also the duration until the event occurred.

The duration is put into a functional relationship with explanatory variables which can reflect both an individual's subjective perception of an innovation[52] just as the individual characteristics of the decision-makers. In contrast to the logistic regression, this approach allows not only to ascertain the adoption probability at a specific point in time but more importantly these probabilities can be determined for each individual *at any point in time*. This enables a more realistic forecast of adoption behaviour. Eventually, by aggregation of the individual probabilities the macro-level can be established hereby illustrating the diffusion process over time.

The process under study (i.e. the adoption process) starts with the market placing of the innovation and ends with the adoption of a sample member at time t_i. The duration of an episode is represented by a random non-negative continuous variable t_i for the ith sample member. This implies that the time-to-event duration is interpreted as the realisation of a random process.[53]

---

[50] Reingannum (1982), Caudil et all (1995), Hannan, Mc Dowell (1984, 1987), Gourlat, Pentecost (2000), Litfin (2000).
[51] Audretsch, Mahmood (1995), Adretsch (1991), Honjo (2000), Kaufman, Wang (2001), Mata et al. (1995).
[52] Litfin (2000).
[53] Logistic regressions and hazard rate models are therefore both stochastic techniques.

As said, the time-to-event duration t_i depends on a number of explanatory variables. These are combined in the vector X_i. The duration of an episode t_i ($t_i \geq 0$) follows a specific distribution that is represented by the **distribution function** $f(t_i)$. The respective **density function** is given by $f(t_i)$. The observation period has the length $(0, T]$.

The following relation between the density function and the **cumulative distribution function** can be established:

$$(2\text{-}1) \qquad F(t_i | X_i) = \Pr(T \leq t_i | X_i) = \int_0^{t_i} f(v_i | X_i) dv$$

and under the assumption that the density function is continuous:

$$(2\text{-}2) \qquad f(t_i | X_i) = F'(t_i | X_i),$$

where

t_i: Episode's duration of the ith member of the sample (i ε I),

X_i: Vector of the causal variables of the ith member (i ε I),

$F(t_i)$: Distribution function of the episode's duration for the ith member (i ε I),

$F'(t_i)$: Derived distribution function of the episode's duration for i^{th} member (i ε I),

$f(t_i)$: Density function of the episode's duration for i^{th} member (i ε I),

$\Pr(T \leq t_i)$ Adoption probability of ith member (i ε I) during the observation period,

$(0, T]$: Length of duration period (here: 0 time of innovation market placing, T time of data collection).

In the context of hazard models, the so-called **survivor function**[54] plays an important rule, too. This function is defined as

$$(2\text{-}3) \qquad S(t_i | X_i) = 1 - F(t_i | X_i) = \Pr(T > t_i | X_i)$$

and represents the *probability that the ith member experiences (i.e. "survives") the point in time t_i*, which is equivalent to the probability that the member has not yet adopted the innovation at this point.

---

[54] Also-called "survival function".

Dependent on the assumed distribution of t_i across all members of the sample, there exists a number of differing survivor functions, which all share one feature: All survivor functions fall monotonously as time proceeds. Translated into the adoption context, this means that the probability of no adoption decreases and the probability of adoption increases with time. Furthermore, the survival probability is *1* for a duration of *0* and *0* for an infinite duration. Yet, the process differs in between these two extremes, whereas explanatory variables can have both, an accelerating and a delaying effect on the survival probability. The following relation we get when time is measured continuously:

$$(2\text{-}4) \quad S(t_i \mid X_i) = 1 - F(t_i \mid X_i) = \int_{t_i}^{\infty} f(v_i \mid X_i) dv$$

The **hazard rate**[55] is defined as

$$(2\text{-}5) \quad h(t_i \mid X_i) = \lim_{\substack{\Delta t \to 0 \\ \Delta t_i > 0}} \frac{1}{\Delta t_i} \Pr(t_i \leq T < t_i + \Delta t_i \mid T \geq t_i, X_i)$$

The aim of the hazard rate is to quantify the conditional probability (i.e. the risk/hazard) that the event "adoption" has already taken place for the ith member at time *t*. As time is a continuous variable, the probability will have the value of *0* at exactly one point in time. For this reason, not a point in time, but a very small time interval $(t_i; t_i + \Delta t_i)$ is observed. The hazard rate function completely describes the probability distribution of the time until an event.

Furthermore, the condition is made that no adoption took place prior to that time interval. Otherwise, the risk of adoption would be redundant. To prevent that the hazard rate is inflated by the dimension of the time interval, the following measures are taken: First of all, only a small time interval is considered and secondly, the probability is adjusted by dividing it by the dimension of the time space Δt_i.[56]

---

[55] As the hazard rate is applied in various scientific fields, there exist various terminologies for it: e.g. transition rate, intensity rate, mortality rate (see Blossfeld, Hamerle, Mayer (1986, p. 31).
[56] Allison (1995).

Henceforth, the hazard rate can be interpreted as the marginal value of the conditional probability that the adoption takes place within the time interval (t_i; $t_i + \Delta t_i$) under the condition that no adoption took place prior to the beginning of the time interval and that the vector X_i is given.

Note that, in contrast to the survivor function, which focuses on non-adoption, the hazard rate focuses on adoption, that is, on the event occurring. Thus, in some sense, the hazard function can be considered as giving the positive side of the information given by the survivor function, That is the higher $S(t)$ is for a given t, the smaller is $h(t)$ and vice versa.[57]

If the ith member "survives" the point in time t_i, then the hazard rate informs approximately about the future process of the probability that the event takes place. The hazard rates can greatly differ in progress. The only restriction is that of non-negative hazard rates. Choosing an alternative formulation for the density function reveals its similarity to the hazard rate,[58]

$$(2\text{-}6) \qquad f(t_i \mid X_i) = \lim_{\substack{\Delta t \to 0 \\ \Delta t_i > 0}} \frac{1}{\Delta t_i} \Pr(t_i \leq T < t_i + \Delta t_i \mid X_i)$$

The only difference between equations (2-5) ("hazard rate") and (2-6) ("density function") lies in the restriction; while in equation (2-6) the probability depends merely on the vector X_i, in equation (2-5) the condition that the adoption has not yet taken place before the t_i holds additionally.

The hazard rate (2-5), the density function (2-1), and the survivor function (2-3) all constitute equivalent forms to describe the continuous probability distribution of the random variable t_i in dependence on X_i. The relation between the function can be derived from the above equation as follows:[59]

$$(2\text{-}7) \qquad h(t_i \mid X_i) = \frac{f(t_i \mid X_i)}{S(t_i \mid X_i)} = \frac{f(t_i \mid X_i)}{1 - F(t_i \mid X_i)}$$

---

[57] The interpretational power of the hazard rate as opposed to the survivor function is thus stronger in Adoption and diffusion theory as in survival analysis.
[58] See Allison (1995), p. 16, Kleinbaum (1995) p. 11.
[59] See Allison (1995), p.16.

Although the process under study is fully described by one of these functions, it should be clear that a distinction between these is useful as every function centres on differing aspects. The hazard rate can be interpreted as the "risk" that an adoption has taken place within the observed time period under the condition that no adoption has yet taken place.[60] Furthermore, the survivor function provides information about the probability that the sample member survives that time period (i.e. that no adoption takes place within the observed duration). For each member of the sample this information exists at any point in time within the observation period.

In order to empirically estimate hazard models for all members of the sample with size I, the duration until the event just as the corresponding explanatory variables for the observation period $(0,T]$ have to be known. In most cases the end of the observation period is pre-determined by the mere fact that random samples are drawn retrospectively. As a consequence, the following **censoring problems**[61] can arise:[62] We talk about **left censored data**, when the event "adoption" has already taken place before the beginning of the observation period. In this case, we only know about the fact that the sample member has already adopted. Whereas left censored data describe events that have taken place before the beginning of the observation period, **right censored data** refer to events that occur after the observation period has ended. In the case of right censored data, one only knows that sample member have not yet adopted the innovation, but we do know at what point in time after the observation period the adoption will take place.

To distinguish between individuals (or firms) who experience the event from those who are censored, we usually use a dichotomous variable δ that indicates the censorship status. Thus, δ_1 is *1* if firm 1 gets the event or is *0* if the observation is censored. The standard assumption made in event history models is that the failure and censoring mechanisms are independent and that censoring is not affected by the covariates.

The problem of censoring is illustrated in figure 2-1: Observations 1 and 3 are seen as complete, as here one can exactly identify the time-to-event duration and the respective explanatory variables. The observations for sample members 2 and 4 are censored. For them, it is impossible to determine the time-to-event duration. With

---

[60] In this way, larger hazards are directly related to shorter survival or earlier adoption.
[61] We have already referred to this phenomenon as "partially missing information".
[62] Blossfeld, Hamerle, Mayer, (1986), p. 72 ff.; Klein Moeschberger (1997), p. 55 ff.

respect to sample member 2, one only knows that at the end of the observation period he still belongs to the group of non-adopters and for member 4 that he already adopted any time before the observation period.

Figure 2-1: Right- and left censored data in ADT

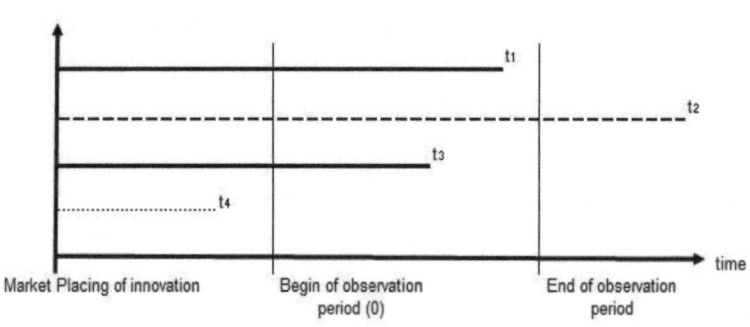

Following Litfin (2000), p. 46

The problem of left censoring can be solved without difficulty in the current context. For this, we let the start of the observation period correspond to the date of the innovation's market placing, so that no adoption can possibly happen before the observation period. This seems easy as all data is collected retrospectively anyway.

The problem of right censoring, on the other hand, cannot be solved that easily. If the goal of the analysis is not only to ex-post examine adoption behaviour and the diffusion process, but also to forecast the diffusion process we face practical obstacles. Logically, it is impossible to move the end of the observation period forward until all observations are complete. Elimination of the censored data is out of the question, as this would lead to distorted estimations.[63] Here, one rather uses all available information about observations in the analysis.

Although, we have no information about the duration for the right censored data, we know that no adoption has yet taken place. This information has to be considered in the maximum likelihood function. Various models have been developed to successfully consider right censored data in hazard models.[64] As mentioned earlier, the integration of censored data is a unique feature of the hazard models.

---

[63] Blossfeld, Hamerle, Mayer (1986), p 72.
[64] Allison (1984), Blossfeld, Hamerle, Mayer, pp. 72-74 (1986) provide a very conclusive overview of the various censoring models.

2.4 Classical Methods for the Analysis of Event History Data

In section 2.2, I provided an introduction into the concepts and terminologies of event history analysis. Subsequently, I established the statistical framework of event history models. I mentioned that comprehension of these ideas is essential in understanding the models that we are to present in this section. Also, I had mentioned that in introducing the classical methods for event history analysis, I would focus on the classical methods used for survival data. As survival analysis occupies a predominant position within event history analysis, these methods can and are, indeed, seen as the classical methods for event history analysis, too.

We can divide the hazard models into 3 well-known categories: non-parametric, parametric and semi-parametric models.[65] To understand what these models try to investigate, I will state, in general terms, **three principal goals of survival analsis**:[66]

Goal 1: To estimate time-to-event for a group of sample members[67]

Goal 2: To compare time-to-event between two or more groups[68]

Goal 3: To assess the relationship of explanatory variables to time-to event

Certainly, goal 3 can be said to have always been of central interest to both survival and adoption- and diffusion modelling.

As for the non-parametric methods, these fulfil mainly descriptive purposes and are generally used to estimate and compare the survivor and/or hazard functions. Both the parametric and semi-parametric approach are regressions of the hazard rate. As opposed to the parametric hazard models, Cox's semi-parametric hazard model merely parametrifies the influence of covariates, but not the time-dependence of the hazard rate. The advantage of this model is the fact that one doesn't need to take assumptions concerning the distribution of the duration.

2.4.1 Non-Parametric Methods

Because these methods do not make any assumptions about the distribution of the process under investigation, *non-parametric methods are often used for ad-hoc*

---

[65] Although we can subsume all methods that we will introduce under the description "hazard models", one often mainly refers to the semi-parametric or parametric approach when this description is used.
[66] Following Kleinbaum (1995), p. 15.
[67] I.e. to estimate and interpret survivor and/or hazard function.
[68] I.e. to compare survivor and/or hazard functions.

examination of survival distribution and first exploratory assessment of variable relations. Moreover, these methods fulfil primarily descriptive purposes.

One obvious first numerical approach to survival data that I initially want to mention is the assessment of the median or mean survival time. In table 2-1, the mean adoption (survival) time of e-sale for a dataset of European enterprises is exhibited as a by-product of the Kaplan-Meier estimate calculations.[69] Additionally, a 95% confidence interval, showing very low variability in data, is calculated for the mean adoption time. The median adoption time cannot be estimated if the number of events is much less than half of the observations studied. This is the case here.

Considering that the observation period was 9 years, the exhibited data clearly tells us that most of the assessed adoption durations lie at the very end of the observation period. Yet, censored data is not reasonably considered here as calculation is done on the basis of observed adoption time implying an adoption time of 9 years for observations that were censored at the end of the observation period. This is why the mean adoption time, as such, provides very limited insight in the data structure and is never used solely.

Table 2-1: Mean adoption time for e-sale

Survival Analysis for esale_t censoring time of online sales adoption til 2002

	Survival Time	Standard Error	95% Confidence Interval	
Mean: (Limited to	8,71 9,00)	,01	(8,69;	8,74)

Source: SPSS output for Kaplan-Meier estimates

In general, literature describes mainly two alternative non-parametric methods both of which are helpful for graphical presentations of the survivor function.[70] The **life table method** is the more traditional method procedure and has been used in the case of large data sets because it needs less computing time and space.[71] However, compared to the Kaplan-Meier estimator[72], the life table method has the disadvantage that the researcher has to define discrete time intervals. Given modern computers, there seems to be no reason anymore to prefer the life table method on

---

[69] Calculation was performed in SPSS.
[70] Blossfeld, Rohwer (2002), p.56.
[71] For a detailed description of life-tables see Blossfeld, Hamerle, Mayer (1986), p. 30.

the basis of computer time or storage space.⁷³ Furthermore, we will see later that *Kaplan-Meier estimates play an important reason in the construction of survival trees.* For these reasons, I will introduce and use the Kaplan-Meier estimator exclusively.

In addition to the Kaplan-Meier estimator, I will present the most common method used to evaluate whether or not the survival distributions for two or more groups are statistically equivalent (compare survival distributions statistically).⁷⁴ We will meet this method, too, in the context of survival trees. This method is a statistic test and is usually called "Mantel's log-rank test" or simply "log-rank test". It has been generalized from the Savage's test (1956) and received its final name from Peto and Peto (1972).⁷⁵

In today's survival analysis Kaplan-Meier estimates⁷⁶ have become the method most widely applied for the assessment and comparison of prognostic classification schemes. The goal of the method is to produce the survivor function without having to specify the distribution of survival times. Kaplan-Meier defines an estimate $\hat{S}$ by

$$(2.8) \qquad \hat{S}(t) = \prod_{j\,:\,t_j \leq t} \frac{n_j - d_j}{n_j}$$

In equation (2.8) there are n observations and $k \leq n$ observations at which failure events are observed to occur; n_j is the number of observations which have not yet failed or are censored; in other words the number of observations at risk at time j and d_j is the number of failures observed at time j.

The Kaplan-Meier estimator is a step function with drops at the observed event times. The size of these drops depends not only on the number of events observed at each event time, but also on the pattern of the censored observations prior to t_j. The vertical axis represents estimated probability of survival (i.e. non-adoption) for a hypothetical cohort.⁷⁷

I estimated the Kaplan-Meier survivor functions for the adoption of 3 e-business technologies by companies as shown in figure 2-2. The respective e-business

---

[72] Kaplan, Meier (1958).
[73] Blossfeld, Rohwer (2002), p.56.
[74] The Breslow and the Tarone test are other tests frequently used.
[75] They called it „Mantel's log-rank test" due to the authors mayor contribution to the development of this test statistic.
[76] Sometimes it is also-called the non-parametric maximum likelihood estimate of the survivor function or product-limit estimate.
[77] Not actual % surviving

technologies were knowledge management software solution, e-sales (i.e. offering products for sale online), e-purchase (purchasing products online). One can see that e-purchase experiences the highest adoption rates. Apart from apparent differences in adoption distributions, we can see that the right censored data, for all three processes, was censored at the end of the observation period (2002).

Although the survival curves can surely be used in this form to compare survival distributions, I have come to the conclusion that the "1-survivor function" representation and the hazard function which both can be established on the basis of Kaplan-Meier estimates, give a better and more intuitive idea of the process in the context of ADT. In the future, I will consequently use the 1-survivor function for first ad-hoc insight in the adoption process under investigation. The reason is the easier interpretability of the survivor function as opposed to the hazard rate.

Figure 2-2: Kaplan-Meier estimate of the survivor functions
For adoption of 3 e-business solutions

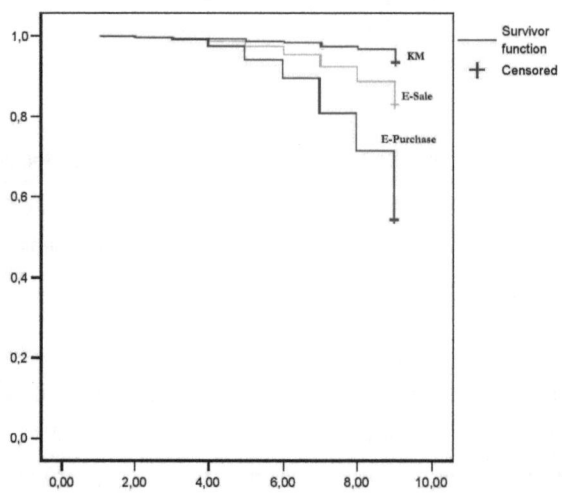

As mentioned, Kaplan-Meier curves are often accompanied by p-values of the **log-rank test** to examine homogeneity across risk strata. This method can also be used to check the adequacy of a parametric event history model.

The log-rank test is an intuitively sensible and straight-forward-to-compute non-parametric test for establishing differences between distributions for two or more groups. The underlying question is whether survival distributions of group a and b are different?

A simple, but often unsatisfactory option would be to compare Kaplan-Meier curves graphically. Although graphical illustrations are useful, it is also important to test the significance of the difference in survival distributions.[78]

Of course, another way of addressing this issue is to construct the upcoming hazard rate regression models for survival time, and to make inferences about the treatment effect by estimating and testing hypotheses about corresponding model parameters. Here, however, a method is presented that can compare survival distributions non-parametrically making the fewest possible assumptions about the survival distribtions.

The procedure allows for censored data and has high power to detect differences when such differences correspond to proportional hazards e.g. $S_1(t) = S_2(t)^\varphi \Rightarrow h_1(t) = \varphi \cdot h_2(t)$.

As an example of the information required for the log-rank test, the typical comparison of two allegedly different groups of subjects ($i = 1,2$) is considered. We denote the ordered observed failure times by $t_{(1)} < t_{(2)} < ... < t_{(k)}$. Then, the log-rank test proceeds by considering the following 2x2 contingency table at each $t(j)$.[79]

Table 2-2: Log-rank 2x2 matrix

Group (i)	Failed at $t(j)$	Total at risk at $t(j)$
1	d_{1j}	n_{1j}
2	d_{2j}	n_{2j}
Total	d_j	n_j

Source: Lecture notes Forster (2004)

Here, for each ordered failure time, $t_{(j)}$ in the entire set of data, the numbers of subjects (d_{ij}) failing (adopting) at that time are shown separately by group (i), followed by the numbers of subjects (n_{ij}) in the risk set at that time, also separately by group.

Under the *null hypothesis*, H_o, of no difference between the groups, all subsets of size $d_{(j)}$ of the $n_{(j)}$ individuals at risk at $t_{(j)}$ are equally likely to comprise the set of

---

[78] Comparison of two survival distributions is particularly important in, for example, clinical studies where groups of patients have been randomised to different treatments and we are required to make inferences about the effect of treatment on survival.
[79] Blossfeld, Hamerle, Mayer (1986), p.47.

failures at this time. Hence, the distribution of $d_{(1j)}$ under the null hypothesis is hypergeometric[80] and has mean and variance,

$$(2\text{-}9) \qquad e_{1j} \equiv E(d_{1j}) = \frac{n_{1j} \cdot d_j}{n_j} \quad \text{(Mean)}$$

and

$$(2\text{-}10) \qquad v_{1j} \equiv Var(d_{1j}) = \frac{n_{1j} \cdot n_{2j} \cdot d_j(n_j - d_j)}{n^2{}_j \cdot (n_j - 1)} \quad \text{(Variance)}.$$

The log-rank test statistics is given by

$$(2\text{-}11) \qquad T = \frac{(\sum_j (d_{1j} - e_{1j}))^2}{\sum_j v_{1j}},$$

which has approximately a $\chi^2{}_1$ distribution under the null hypothesis, provided that k is not too small. Another way of expressing this test statistic is

$$(2\text{-}12) \qquad T = \frac{(O - E)^2}{Var(O - E)},$$

where $O = \sum_j d_{1j}$ is the observed number of deaths in group 1 and $E = \sum_j e_{1j}$ is the expected number under the null hypothesis of no difference between the groups.[81]

When we state that two Kaplan-Meier curves are "statistically equivalent," we mean that, based on a testing procedure that compares the two curves in some "overall sense" we do not have evidence to indicate that the true (population) survival curves are different.

The log-rank test is a large sample chi-square test[82] that uses as its test criterion a statistic that provides an overall comparison of the Kaplan-Meier curves being compared. This (log-rank) statistic, like many other statistics used in other kinds of chi-square tests, makes use of an observed versus expected cell counts over categories of outcomes.[83]

As we have seen, the Kaplan-Meier estimate can be used to depict the survivor function hereby providing a first rough insight into the survival distribution. Besides,

---

[80] Forster (2004), p. 2.
[81] Kleinbaum (1995), p.61.
[82] Kleinbaum (1995), p. 58.
[83] Lee (1992), p. 104.

with the help of the log-rank test, as a univariate analytical tool, it can detect differences in adoption time caused by a single factor.

Usually, however, one is interested in the impact that multiple covariates have on the adoption distribution. This brings me to the hazard rate regressions, both semi-parametric and parametric.

2.4.2 Parametric Methods

There a two general answers to the question of how we estimate the survival or hazard function from the collected event history data. The first one assumes no specific knowledge of the survival distribution and hence is non-parametric, or distribution free and the second one does assume to have this specific knowledge and hence is parametric.

A number of possible distributions for the density function have been suggested for the hazard models. Among these, the exponential-, Weibull-, lognormal-, log-logistic-, and Gombertz distribution.

I will only introduce the **exponential distribution** which is the simplest one. For this reason, this distribution assumption is frequently used as reference model for other more complex models. In fact, this distribution assumption is an essential argument in the semi-parametric Cox model and will therefore help to understand it.

The density function, survivor function and hazard rate under the exponential duration distribution assumption are given by:[84]

(2-13) $f(t_i | X_i) = \lambda_i \cdot e^{-\lambda_i \cdot t_i}$,

(2-14) $S(t_i | X_i) = \exp(-(\lambda_i \cdot t_i)^\alpha)$,

(2-15) $h(t_i | X_i) = \lambda_i$,

where

λ_i: Parameter with $\lambda_i = e^{-\beta_0 - x_i \cdot \beta}$,

X_i: Vector of covariates of the ith member of the sample, and

β: Vector of coefficients.

Consequently, the exponential model is fully determined by the parameter λ_i. One can see that the hazard rate in this model remains constant over time, which implies that the probability of adoption is independent of the observed duration. This restriction seems implausible as the information about the innovation first has to spread across the social system. Naturally, one should expect a rising adoption probability as time proceeds. The unknown model parameters are obtained by the maximum likelihood estimation.

Even so, a constant adoption probability over time does not mean that there prevails a constant probability for all members. Individual attributes and attitudes vary among all sample members. These individual attributes and attitudes are set in a direct relation to the parameter λ_i through a log-linear model. Differing expressions of the causal vector X_i result in different adoption probabilities for the sample members.

Differing hazard rates have been illustrated in figure 2-3. For the parameter λ_i the values 1 and 2 were taken.[85] The figure underlines that the hazard rates for all members are time-independent when time-constant explanatory variables are considered. Besides, we can see that the hazard rate of one sample member is a multiple of the other member's rate. Thus, the sample member's hazard rates are proportional to each other.

Figure 2-3: Exponential distribution of survivor function

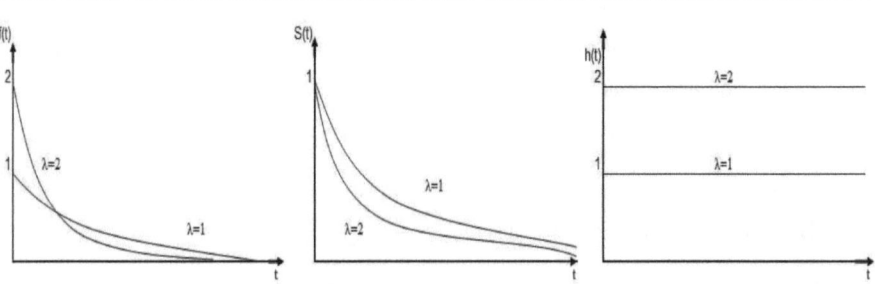

Source: Blossfeld, Hamerle, Mayer (1986), p. 35

Formally, we can express this proportionality as follows:

(2-16) $\qquad \dfrac{h(t \mid X_j)}{h(t \mid X_i)} = \dfrac{\exp(\beta_{0j} + X'_j \cdot \beta_j)}{\exp(\beta_{0i} + X'_i \cdot \beta_i)}$ for $i \neq j$.

---

[84] Following Blossfeld Hamerle, Mayer p. 52.

This equation demonstrates that the sample member's hazard rate is independent of time. Hence, the exponential model fulfils the characteristics of Cox proportional hazard model[86]. Formally, for any time t, $h_i(t)/h_j(t) = c$, where i and j refer to the distinct individuals and c may depend on explanatory variable but *not on time*.

A widespread criticism of parametric hazard rate models is the inherent necessity to decide how the hazard rate depends on time. In reality, there is usually little information on which to base such choice. Moreover, if the hazard is believed to be non-monotonic, it may be difficult to find a model with the appropriate shape.

Much experience with these models suggest that the coefficient estimates are not terribly sensitive to the choice of the hazard function, but one can never be sure what will happen in any particular case. Another problem is the consideration of explanatory variables that change over time; while it is possible to develop fully parametric models that include time-varying explanatory variables,[87] estimation of these models is somewhat cumbersome. For these reasons the fully parametric hazard rate models are rarely used in both survival analysis and ADT.

Both of the above described problems were solved in 1972 when David Cox, a British statistician, published a paper entitled "Regression Analysis and Life Table" in which he proposed a model and an estimation method that have become extremely popular since. The Cox model has become the event history model primarily used not only in survival analysis[88] but in all other scientific fields, too.[89]

2.4.3 Semi- Parametric Methods

Commonly referred to as the "proportional hazard model", Cox's model (1972) is a generalization of the parametric models we have considered above. I will limit myself to explaining merely the main concept of this model as certain inherent mathematical issues can become rather complex.[90] The general idea, however, is surprisingly simple.

---

[85] See Blossfeld, Hamerle, Mayer (1986), p. 52 ff.
[86] Cox (1972).
[87] Tuma (1979).
[88] Radespiel-Tröger et al. (2003), p. 3.
[89] The Cox model is one of the most frequently cited model in the entire scientific literature.
[90] There has emerged an inexhaustible number of descriptions and extensions to this models and its inherent features since its publication in 1972.

No surprise, it is called the proportional hazard model because for any two sample members at any point in time the ratio of their hazard is a constant. Instead of making assumptions directly on survival times, Cox proposed to specify the hazard function, which means that we do not have to worry about the numerical value the hazard rate takes.

The most widely used expression for the proportional hazard function is

(2-17) $h(t; X_{it}, \beta) = h_0(t)\exp(X_{it} \cdot \beta)$,

where $h_0(t)$ is the baseline hazard, X is a vector of explanatory variables which may incorporate the rank, stock or order effects, and β is a vector of parameters to be estimated.

In contrast to the parametric approach, the Cox model leaves the baseline hazard $h_0(t)$ unspecified. This model is semi-parametric because while the baseline can take any form, the covariates enter the model linearly. Remarkably, even though the baseline hazard is unspecified, the Cox model can still be estimated by the method of partial likelihood developed by Cox in the same paper in which he introduced the Cox model.[91] Although the resulting estimates are not as efficient as maximum-likelihood estimates,[92] for a correctly specified parametric hazard regression model, not having to make arbitrary, and possibly incorrect, assumptions about the form of the baseline hazard is a compensating virtue of Cox's specification.

The output of the Cox model has many of the same components as the output from a multiple regression model.

Again, I have taken an example from e-purchase adoption among European companies to illustrate what results one gets from the Cox model.[93] I have consciously accentuated the independent variable z01b_8[94]. This variable has resulted in the highest positive hazard coefficient, so it is worthwhile taking a deeper look at this variable.

---

[91] The ability to estimate the model by the method of partial likelihood widely seen as Cox's mayor contribution.
[92] Estimation of this model can be performed through full likelihood, too, but this is very difficult and complex.
[93] The whole model's significance was verified via the log-likelihood output for the model.
[94] Was labelled as dummy variable; company section ICT services (0=NO, 1= Yes).

Table 2-3: Coefficients from semi-parametric Cox Model

Equation Variables

	B	SE	Wald	df	Signifikanz	Exp(B)
p012	,578	,232	6,194	1	,013	1,782
p013	-,281	,231	1,483	1	,223	,755
p014	-,342	,250	1,879	1	,170	,710
z01b_8	1,162	,145	64,479	1	,000	3,197
z01b_9	,482	,129	13,926	1	,000	1,619
z01b_10	,299	,134	4,953	1	,026	1,348
z01b_11				0ª		
g01q_1	-,057	,075	,587	1	,444	,944
g01q_2	-,057	,072	,619	1	,431	,945
g01q_3	,005	,069	,006	1	,938	1,005
g01q_4				0ª		
g02_1	,237	,303	,613	1	,433	1,268
g02_2	,423	,304	1,933	1	,164	1,526
g06_1	,097	,249	,154	1	,695	1,102
g06_2	,217	,248	,766	1	,381	1,243
g06_3	,136	,251	,292	1	,589	1,146
g06_4	,239	,252	,900	1	,343	1,270
g11_1	,277	,073	14,363	1	,000	1,319
g11_2	,326	,068	23,130	1	,000	1,385
g11_3	,260	,079	10,871	1	,001	1,297
g11_4	,270	,070	14,865	1	,000	1,310
g11_5	,195	,053	13,560	1	,000	1,216
g11_6				0ª		

Source: SPPS output; Cox Regression

Without going to much into the calculation pattern, the interpretation of this coefficient may be phrased as follows: Companies that come from the section of ICT services exhibit the highest risk of e-purchase adoption across time. Out of all variables, this company characteristic proves to be the best predictor in terms of telling whether a company will adopt early within the diffusion process.[95] The log-hazard ratio is 3.197 implying that the risk of adoption for companies coming from ICT services is approximately twice that of those from business services (z01b_9). The whole model's high significance was verified via the log-likelihood output for the model ($p<0.001$).

This model is the most widely used event history model in investigating the relationship between adoption time response variables and covariates. As opposed to parametric approaches, it has been used to analyse the determinants of technology diffusion in a number publications.[96]

---

[95] Statistical significance was achieved.
[96] Gourlay, Pentecost (2000), Hannan, McDowell (1987), Ingham, Thompson (1993), Saloner, Shepard (1995).

3 Presentation and Analysis of the Survival Tree Method

During the past few years several non-parametric alternatives to classical hazard models have appeared in survival literature. These methods mostly extend techniques that are well known from regression analysis to the analysis of event data. Among them, the so-called HARE[97] and survival tree method have been discussed most centrally.[98] Whereas, HARE are based on (polynomial) splines, survival trees are based on partition trees.[99] Both methods have proven to resemble useful contributions to survival analysis.[100]

Survival trees are a non-parametric method that is able to assess multiple covariate *effects* on survival time. But not only in this they differ from classical non-parametric methods; the method has a number of persuasive advantages that explain its growing acceptance in survival analysis. Yet, outside of survival analysis, applications of it have materialized only in life history analysis[101] and food science[102].

I will use this chapter to describe and discuss the inherent features of survival trees before investigating what insight this method can offer to ADT.

As the formation of prognostic subgroups has often been quoted as the method's main advantage, it should come as no surprise that they have emerged from the field of survival analysis. Here, the identification of risk groups of patients defined by values of certain prognostic factors has always been of extraordinary importance. A medical decision (e.g. the application of correct treatment) may essentially depend on a patient's correct prognosis.

The basic idea of survival trees is to split the dataset into homogeneous groups or clusters with respect to the dependent variable based on the best set of predictors, that is, into subgroups with similar prognosis. The resulting structure is depicted graphically in the form of a tree.

Although probably not obvious at first glance, the formation of prognostic (risk) groups is of great importance in the field of economics, too. In innovation marketing,

---

[97] Kooperberg, C. Stone, C, Truong, Y. (1995). HARE stands for hazard regressions. The denomination can be misleading.
[98] Keles, Segal (2002).
[99] It may be interesting to know that neural networks have been extended to censored event data, too, see Faraggi, Simon (1995) for more details.
[100] Intrator, Kooperberg (1995).
[101] Rose, Pallara (1997).
[102] Evans et al. (2004).

for instance, target groups need to be recognized and treated adequately to achieve the critical mass and enhance overall product growth.

In general, tree-based methods involve three major steps: growing the saturated tree, pruning a sequence of optimal trees and selecting the final tree from the sequence of optimal trees.

The statistical underpinnings of this procedure were first detailed in the monograph by Breiman, Friedman, Olshen and Stone (1984) who coined the phrase "classification and regression trees" or "CART ^{TM}"[103]. In few words, *survival trees are nothing more than the extension of the CART^{TM} method to event history data.*[104] Hence, I will start the chapter by developing the concepts and working patterns of the CART^{TM} method. We will see that as the type of dependent variable differs between the CART^{TM} and survival tree method, the notion of homogeneity and thus the appropriate splitting criterion differs, too.

As I developed the argument that models from survival analysis are likely to prove interesting for the modelling of ADT, the investigation of the model's usefulness for this area is a natural step.

3.1   Review of CART^{TM}

In contrast to classical regression and classification models the non-parametric CART^{TM} method does not require a specified model structure. Rather than fitting a model to the sample data, a tree structure is generated by dividing the sample recursively into a number of groups, each division being chosen so as to maximize some measure of the difference in the response variable for the resulting binary subgroup.

In the **first step** the dataset is recursively split (i.e. partitioned) into subgroups until a so-called "saturated tree" is found. The root node of a tree contains the sample of subjects from which the tree is grown.

The criterion that each split has to comply with is that it splits each node into those two sub-groups that are most homogeneous in terms of predicting the binary dependent variable.

---

[103] The TM (Trade mark) has to be used as the method has been commercialized in the respective software package: CART^{TM}.

In the context of ADT, the sample is recursively split into groups that are most homogeneous in predicting whether a company adopted a certain technology or not. In other words, complete homogeneity means that a node contains either only adopters or non-adopters.

Predictor variables can be ordinal, (continuous) or nominal. The number of possible splits varies with the type of predictor variable. Usually, complete homogeneity is an ideal that is rarely realized. Thus, *the numerical objective of partitioning is to make the contents of the nodes as homogeneous as possible.* A number of methods have been proposed to assess the extent of node homogeneity for each node and to choose the best split[105]. In general, splitting rules are based on node impurity measures such as the Gini index, the Bayes rule or the entropy function.[106] Most frequently, the entropy function is used due to a number of desirable properties.[107]

Any of the p covariates $x_1,...x_p$ (with $j=1...p$) is qualified for producing splits. If x_j is ordinal, then we split the population according to whether $x_j < c$ for some constant c. If x_j is a nominal variable that takes its values in $A=\{c_1,...,c_k\}$, then a split is based on whether $x_j \in B$, where B is some subset of A. For a choice of c (or B depending on the type of x_j) one assigns the *i-th* subject to the left node (τ_L) if $x_{ji}<c$ or $x_{ji} \in B$ and to the right daughter (τ_R) node otherwise.

Thus, for each split, the entire set of available predictor variables is considered in order to determine which one maximizes the homogeneity of the following two daughter nodes.

The *goodness of a split* is defined by the decrease in impurity[108]

(3-1) $\quad \Delta I(s,\tau) = i(\tau) - P\{\tau_L\}i(\tau_L) - P\{\tau_R\}i(\tau_R)$.

Eventually, the split with the highest reduction in impurity is chosen.

This process is continued until the nodes are completely homogeneous and cannot be split any further. Those subsets that are not split anymore are called "terminal nodes".

The resulting saturated tree τ_0 is generally very overfit. In other words, it follows every idiosyncrasy in the learning dataset, many of which are unlikely to occur in a

---

[104] Likewise are hazard regressions (HARE) the extension of multivariate adaptive splines (MARS).
[105] Breiman et al. (1984), p. 93-129.
[106] Breiman et al. additionally proposed the twoing rule which is not related to a node impurity measure.
[107] Zhang, Singer (1999); p.30, Köllinger, Schade (2004).

future independent group of sample members. Therefore, we prune the saturated tree T_0, in the **second step**, to find the optimal tree which should be a nested subtree of T_0. The method of "cost-complexity pruning"[109] is usually used to generate a sequence of optimal subtrees $T_{1...x}$, each of which is candidate for the appropriately-fit final optimal tree.

As component of the cost-complexity algorithm, a "classification performance" of the entire tree T needs to be ascertained. This classification performance of the entire tree is given by the sum of the quality of its terminal nodes:[110]

(3-2) $\qquad R(T) = \sum_{\tau \in \tilde{T}} r(\tau) \cdot P(\tau) = \sum_{\tau \in \tilde{T}} R(\tau),$

where

$r(\tau)$: Expected costs given that a subject falls into node τ,[111]

$P(\tau)$: The probability of a subject to fall into the node τ,

$\tilde{\tau}$: Terminal node, and

$R(\tau)$: Misclassification costs for node τ where $\tau \in \tilde{T}$.

For any subtree $T_i < T_0$, its complexity is defined as $|\tilde{T}|$, the number of terminal nodes in T_i. Let $\alpha (\geq 0)$ be a real number called the complexity parameter and define the cost-complexity of the entire tree as

(3-3) $\qquad R_\alpha(T) = R(T) + \alpha |\tilde{T}|.$

For any value of $\alpha (\geq 0)$, *there is a unique smallest subtree of* T_0 *that minimizes* $R_\alpha(T)$.[112] Thus, by gradually increasing α, a sequence of nested optimal subtrees of T_0 can be constructed by pruning off the weakest branches at each threshold level

---

[108] Breiman et al. (1984), p. 25.
[109] Breiman et al (1984), pp. 66-71.
[110] For description of the quality of a node or the entire tree the expressions prediction error, cost or resubstitution estimate of the misclassification cost are used interchangeably. We have observed a great of denominations that create unnecessary complexity in the area. Although, the expressions we use do not correspond fully to the ones used by Breiman et al (1983), we think that they are easier to understand
[111] For a detailed description of the assessment of $r(\tau)$ and the class assignment rule, on which it depends, see Breiman et al. pp. 32-34.
[112] The formal proof is given in Breiman et al. (1985), chapter 10.

of α. Therefore, α is a measure of how much additional accuracy a split must add to the entire tree to warrant the additional complexity.[113]

Eventually, the final tree must be selected from the pruning sequence in the **third step**. The classification performance $R(T)$ as specified in (3-2) is obviously biased and results in severe over-fitting. The solution lies in finding an honest estimate for the true classification performance and selecting the subtree that minimizes the estimated true misclassification costs. This is usually done with an independent test sample, "boot-strapping", or "cross-validation"[114,115].

Cross-validation and boot-strapping are both methods for estimating generalization error based on "resampling"[116]. In k-fold cross-validation, one divides the dataset into k subsets of, approximately, equal size. You train the net k times, each time leaving out one of the subsets from the training, but using only the omitted subset to compute whatever error criterion interests you. If k equals the sample size this is called "leave-one-out cross-validation"[117].

In the simplest form of boot-strapping, instead of repeatedly analyzing subsets of the data, one repeatedly analyzes subsamples of the data. Each subsample is a random sample with replacement from the full sample.

I calculated a classification and regression tree for e-purchase adoption that is depicted in figure 3-1.[118] As the central intent, here, is to give the reader an idea about the appearance of a classification and regression tree, I have deliberately not depicted all nodes for higher graphical convenience.

---

[113] I.e. the cost-complexity measure controls the trade-off between the size of the tree and how well the tree fits the data
[114] Breiman et al (1984), pp. 75-78
[115] Usually X-fold cross-validation is used for classification and regression trees, Gordon, Olshen (1985), p. 4
[116] Efron, Tibshirani (1993)
[117] Leave-one-out cross-validation is easily confused with jack-knifing. Both involve omitting each training case in turn and retraining the network on the remaining subset. But cross-validation is used to estimate generalization error, while the jackknife is used to estimate the bias of a statistic.
[118] The tree was calculated in R with the Rpart module and later re-built for higher graphical convenience. See Appendix 6.1 for the original graphical output from R. Cross-validation was done using the 0-SE rule, which will be described in detail for the survival tree method.

Figure 3-1: Classification and regression tree for e-purchase adoption

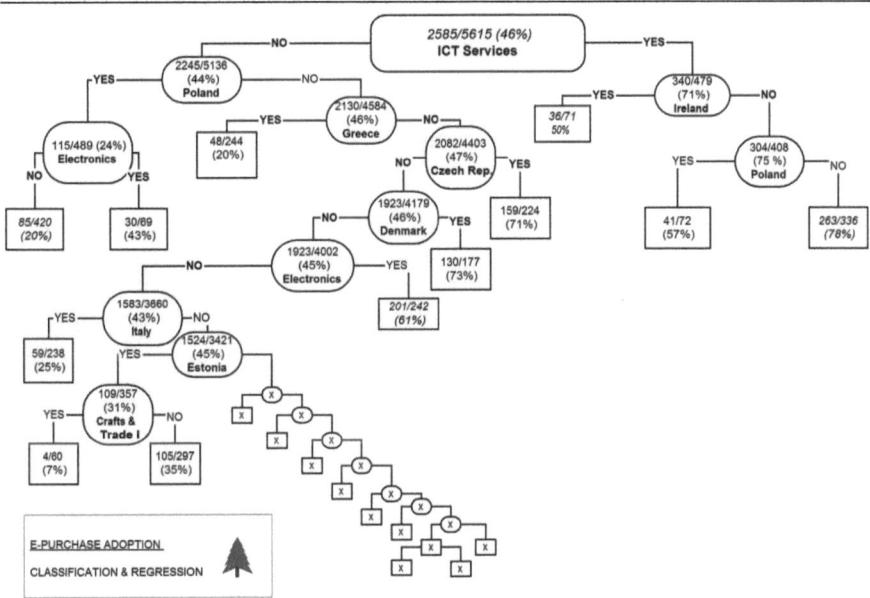

The nodes contain information about the ratio of adopters to total node observations, the resulting adoption probability and the splitting criterion on which the respective node is split into two. The highest adoption probability (78%) is detected in the subgroup including ICT service companies from any country (in the dataset) other than Ireland and Poland. The lowest adoption probability (7%), likewise, is detected in companies from the crafts & trade sector in Estonia.

It is interesting to see that the same criticism I enacted on logistic regressions as being static in nature and ergo taking merely a snapshot of a process whose outcome it should predict, can be equally directed to the CART^{TM} method in the ADT context.

3.2 Principle Framework & Mechanics of Survival Trees

By now the vast majority of publications that have come up in literature deal with survival trees capable of investigating univariate event history data, which one may then call accordingly "univariate survival trees". Most recently, so-called "multivariate survival trees" have been developed, too. [119] These allow for investigation of multivariate event history data time data and thus belong to the multiple risk type of model.[120] In this way, one would subordinate univariate survival trees under single risk models.

As we have identified the adoption process as one with a final absorbing event, univariate survival trees must consequently be the appropriate method to investigate this process.

Hence, in this thesis I will discuss univariate survival trees exclusively. Besides, I will not discuss the integration of time-dependent variables, although possible, here.

Survival trees are the extension of the CART™ method to the analysis of censored event data. As such, the rules of the game are essentially the same as in the CART™ method.

First, a comparable "node impurity" is needed in tree growing; that is, one must define a partitioning criterion by which on node is recursively split into two. **Second**, to guide tree pruning, an analogous "cost complexity" needs to be formulated so that one can choose a right sized tree. And **third**, one needs to identify a procedure that selects the final tree from the sequence of nested optimal subtrees. We will see, too, that alternative proposals have come up that do not follow this classical procedure.

In survival analysis, it is frequently of interest to determine which variables affect the survival distribution and whether the effect is valid across all individuals or within subsets. Statistically, these questions are often posed as variable selection and detection interpretations. Survival trees are used as a tool for revealing structure in the data. It is more sensible to see the survival tree method as the extension of the CART™ method to event history data rather than seeing the two methods as being distinct. This is underlined by the fact that survival trees have managed to preserve the CART™ inherent features that have made them an attractive alternative or complement to other methods.

---

[119] Zhang (1998), Su, X. Fan, J. (2001), Su, X. Fan, J. (2004).
[120] See section 2.2.

Likewise, both methods' attractiveness lies in the straight forward result interpretations and their ability to analyse complex nonlinear data sets with many variables by effectively reducing the dimensionality.

3.3 Splitting, Pruning, Tree-Selection & Alternative Proposals for Survival Trees

On the whole, the extension of CART™ to censored survival data can be distinguished by two distinct splitting approaches.[121] Splitting is undoubtedly the most important building block in the tree methodology. In the survival tree methodology this is even more apparent as the splitting criterion is sometimes used for the pruning as well.

The first splitting approach uses a statistic that determines within node homogeneity. Here, the criterion is how similar the survival experiences of observations are in each node. The second approach is based on separation measures. The main ingredient is now a test statistic that distinguishes between survival experiences.

I will try to treat each of the proposals independently. In the survival tree methodology, a greater variety of splitting, pruning and final tree selection proposals than in the CART™ method have come up. Although not all combinations are appropriate, authors have begun to mix splitting rules with pruning rules, in particular.[122] Besides, not all authors use this three step procedure. Some authors simply stop after the tree is grown and interpret the results referring mainly to the upper splits. Others apply a pruning algorithm but fail to provide reasons on what basis the final tree is chosen. This explains why we cannot establish a coherent treatment of the author-specific approaches for all 3 steps.

I have identified the principle splitting, pruning, and final tree selection rules. Even though not all authors could be named explicitly, a rather conclusive overview is offered.

---

[121] This distinction may sometimes demand in-depth understanding of the method and thus may be difficult to understand at first glance.
[122] Singer, Zhang (1999), p. 100.

3.3.1 Splitting – Growing the Saturated Tree

In classifying binary outcome in the CART^{TM} method, the impact of using different splitting criterion is relatively minor. However, the impact appears to be greater for the analysis of censored data. The choice of an appropriate splitting algorithm for survival trees is not obvious.

What would be an appropriate measure of node impurity in the context of censored data?

Survival trees partition the dataset into subpopulations in which subjects are at about the same risk (of failure/adoption).[123] The intent is to divide subjects into two groups so that the difference in the incidence rates (of the failures/adoptions) in the two groups is as extreme as possible. In that way, one makes the incidence rate as high as possible in one group and as low as possible in the other group.

Consequently, *a node can be seen as pure if all failures in the node occurred at the same time.* In such circumstances the shape of the Kaplan-Meier curve within the node has three possibilities, as depicted in figure 3-2.

---

[123] In clinical studies one often refers to this as "similar within –node prognosis".

Figure 3-2: Three possible Kaplan-Meier curves for a homogeneous node

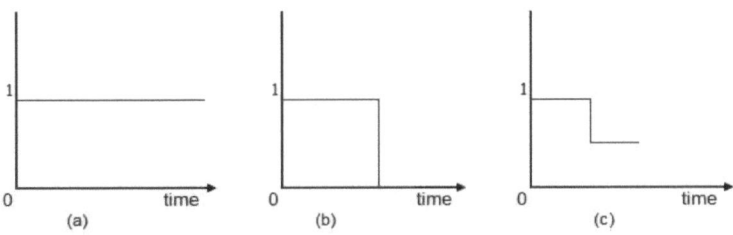

(a) All observations were censored; (b) All failures occurred at the same time and there were no censored observations afterwards; (c) All failures occurred at the same time, followed by censored times. Following: Singer, Zhang (1999), p. 94

Let P be the collection of all such Kaplan Meier curves. One way to judge the node impurity is to see how far the within node Kaplan-Meier curve deviates from any of the curves in P. To this end, one first needs to define a distance between the two Kaplan-Meier curves. This concept brings me to the measures of within node homogeneity.

3.3.1.1 Measures of Within Node Homogeneity

Splitting based on within node homogeneity is conceptually easier as it allows for inheritance of all subsequent CART™ methodology; measures defined are all sub-additive allowing comparison between subtrees. The idea, analogous to CART™, involves minimizing within node variability (i.e. impurity in survival times).

Gordon and Olshen (1985) made the first attempt to adapt the idea of recursive partitioning to cover censored event history data.

They suggested the so-called L^p Wasserstein metrics d_p (.,.) as the measure of discrepancy between the two survivor functions. Graphically, when $p = 1$, the Wasserstein metrics $d_1(S_1, S_2)$ between two Kaplan-Meier curves S_1 and S_2 is the shaded area in figure 3-3. In general, d_p (.,.) is defined as follows:

Let F_1 and F_2 be two distribution functions. The L^p Wasserstein distance between F_1 and F_2 is

(3-4) $$\left[\int_0^1 |F_1^{-1}(u) - F_2^{-1}(u)|^p \, du\right]^{1/p},$$

where $F_i^{-1}(u) = \min\{t : F_i(t) \geq u\}, i = 1,2$.

Figure 3-3: The L^1 Wasserstein distance between two Kaplan-Meier curves

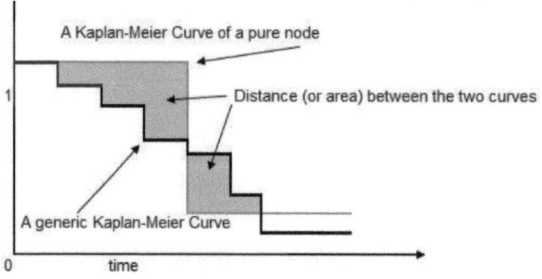

Source: Zhang, Singer (1999), p. 94

Now, let us take $F_1(t) = 1 - S_1(t)$ and $F_2(t) = 1 - S_2(t)$. Note that F_1 and F_2 have all properties of a distribution function except that they may not approach 1 at the right end, which occurs when the longest observed time is censored (see e.g., figure 2-1 a, c).

Formally,

(3-5) $\quad \lim_{t \to \infty} F_1(t) = m_1 \leq 1$ and $\lim_{t \to \infty} F_2(t) = m_1 \leq 1$.

Such F_1 and F_2 are called improper distribution functions. If we can generalize the distance metrics into proper distribution functions, then we can define the distance between two Kaplan-Meier curves as that between the respectively flipped improper distribution functions. Indeed, the L^p Wasserstein distance between S_1 and S_2 can be defined as

(3-6) $\quad \left[\int_0^m |F_1^{-1}(u) - F_2^{-1}(u)|^p \, du \right]^{1/p}$, where the upper limit of the integral m is the minimum of m_1 and m_2. To avoid technicalities, this definition is slightly simpler than the original version of Gordon and Olshen.[124]

We are now ready to define the node impurity. If a node is pure, the corresponding Kaplan-Meier curve should be one of the three curves in figure 3-2. Respective comparisons reveal the degree of node impurity. In formal terms, the impurity of node τ is defined as

(3-7) $\quad i(\tau) = \min_{\delta s \in P} d_p(S_\tau, \delta s),$

where S_τ is the Kaplan-Meier curve within node τ, and the minimization $\min_{\delta s \in P}$ means that S_τ is compared with its best match among the curves of the form depicted in figure 3-2.

On the whole, the numerical implementation of (above) is not a straight forward task, although it is clear, in a theoretical sense, thanks to the fact that the distance is a convex function. When $p=1$, the impurity in (3-7) can be viewed as the deviation of survival times toward their median. Other than the theoretical generality, there is no loss for us to choose p equal to either 1 or 2. After the preparation above, we can divide a node into two as follows. First, we compute the Kaplan-Meier curves for each daughter node. Then, we calculate the node impurities from (3-7). A desirable split can be characterized as the one that results in the smallest weighted impurity. Just as for CART™, the goodness of split is calculated for all available predictor variables and the best predictor, which is the one with the highest $\Delta I(s,t)$, is selected. Formally,

(3.8) $\Delta I(s,\tau) = i(\tau) - P\{\tau_L\}i(\tau_L) - P\{\tau_R\}i\{\tau_R\}.$

---

[124] Simplified description follows Singer, Zhang (1999).

Likewise, the recursive partitioning process continues until the tree is saturated in the sense that the offspring nodes, subject to further division, cannot be split any further (e.g. when there is perfect homogeneity in the node).

Within the classical regression arguments, one can say that Gordon and Olshen suggest replacing least squares with the minimum horizontal or vertical distance between the product-limit estimator (Kaplan-Meier curves) for the node and any step function.

Two likelihood based splitting criteria have been proposed in the survival tree literature. These two and the one that will be presented afterwards employ substantial ideas of the Cox model (1972) and are equally straight forward in their general ideas.

Davis and Anderson (1989) assume that the survivor function within any node is an exponential function with a constant hazard. Within each node, the likelihood function can be easily obtained. Under their assumption, the maximum of the log-likelihood in the node τ is

$$(3\text{-}9) \quad l(\tau) = \sum_{i \in \tau} \delta_i \left[\log(\hat{\lambda}_\tau) - 1 \right],$$

where $\hat{\lambda}_\tau$ is the hazard estimate. They select the split that maximises $l(\tau_L) + l(\tau_R)$.

Likewise, the splitting criterion of **LeBlanc and Crowley (1992)** is based on the assumption that the hazard functions in two daughter nodes are proportional but unknown.[125,126] Accordingly, they develop the idea that the hazard $\lambda(t \mid x_i)$ at time t for individual i with covariates x_i is the product of a baseline hazard that depends only on time and a structural component that, further, depends on the individual through its covariates $\lambda(t)\theta(x_i)$. Furthermore, they base the within-node homogeneity on a one-step deviance residual. The (full likelihood) deviance residual for node τ is

$$(3\text{-}10) \quad R(\tau) = 2[l_\tau \ (saturated) - l_\tau \ (\hat{\theta}_\tau)],$$

where $l_\tau \ (saturated)$ is the log-likelihood for the saturated model that allows one

---

[125] LeBlanc, Crowley (1992), p. 423.
[126] From a methodological point of view, it is useful to know how parametric ideas can be adopted in the non parametric framework. For a detailed description of this see LeBlanc, Crowley (1992), pp. 412-414 and Singer, Zhang (1999), p. 98-99.

parameter for each individual i and $l_i(\hat{\theta}_\tau)$ is the maximized likelihood estimate for the present tree based on the proportional hazard model. LeBlanc and Crowley use a one-step estimate of the deviance based on the Breslow estimate (1972)[127,128] to determine the one-step estimate of $\hat{\theta}_\tau$.[129]

The deviance residual for an individual i in node τ is calculated as

(3-11) $\quad d_i = 2[\delta_i \cdot \log(\dfrac{\delta_i}{\Lambda_0(t_i)\hat{\theta}_\tau}) - (\delta_i - \Lambda_0(t_i)\hat{\theta}_\tau)]$

and the baseline cumulative hazard at node n, given by the Breslow estimate[130], as

(3-12) $\quad \Lambda_0(t) = \dfrac{\sum_{i:t_i \leq t} \delta_i}{(\sum_{i \in \tau} \sum_{i:t_i \geq t} 1)}$.

They show that this impurity measure can be taken as the number of observed failures for individual i minus an estimate of the expected number of failures under the assumption of the tree-structured proportional hazard model.[131] The quality of a node is then given by

(3-13) $\quad R(\tau) = \dfrac{1}{N} \sum_{i \in S\tau} \left[\delta_i \cdot \log\left(\dfrac{\delta_i}{\Lambda_0(t_i)\hat{\theta}_\tau}\right) - (\delta_i - \Lambda_0(t_i)\hat{\theta}_\tau) \right]$.

As usual, all possible splits for each of the covariates are evaluated and *the variable and split point resulting in the greatest reduction in impurity is chosen.*[132]

---

[127] LeBlanc, Crowley (1992), p. 414.
[128] Here, the Breslow estimate corresponds to the Nelson (1969) cumulative hazard estimator.
[129] This estimate is one for all individuals.
[130] Here, we considered the fact that $\hat{\theta}_\tau = 1$.
[131] Intrator et al. (1995), p. 6.
[132] LeBlanc, Crowley (1992), p. 414.

Analogous to the CART™ method the goodness of split s at node τ is given by the reduction of impurity which is, here, ascertained on the basis of the one-step deviance residual

(3-14) $\quad \Delta I(s,\tau) = i(\tau) - P\{\tau_L\}i(\tau_L) - P\{\tau_R\}i\{\tau_R\} = \Delta R(s,\tau) = R(\tau) - [R(\tau_R) + R(\tau_L)].$

Zhang (1995) introduced a totally different concept. He argues that a homogeneous node should consist of subjects whose observed failure times are close and who are mostly censored or mostly uncensored. That is, the impurity of a node reflects the homogeneity in both observed time and the proportion of censoring.

He uses the following impurity criterion:

(3-15) $\quad i(\tau) = w_t \cdot i_t(\tau) + w_\delta \cdot i_\delta(\tau),$

where

$w_t, w_\delta:$ pre-specified weights[133], and

$i_t(\tau), i_\delta(\tau):$ Impurities of node τ for observed time and censoring.

With

(3-16) $\quad i_t(\tau) = \sum_{node\cdot\tau}(t_i - \bar{t}(\tau))^2 / \sum t_i^2,$

and

(3-17) $\quad i_\delta(\tau) = -v_\tau \cdot \log(v_\tau) - (1 - v_\tau) \cdot \log(1 - p_\tau),$

where $\bar{t}(\tau)$ is the average of t (observed time) in node τ and the sum in the numerator is over the subjects in node τ, whereas the sum of the denominator is either over the subjects in node τ or over the entire study population. The denominator is introduced to normalize the scaling. When the sum is over node τ, it is called adaptive normalization (*AN*). On the other hand, when the sum is over the entire sample, it is called global normalization (*GN*). Finally v_τ is the proportion of censoring in node τ.

---

[133] Zhang (1995a) explored the effect of various weights and found equal weights (1:1) to be a reasonable choice.

In order to extend the CART™ to survival data, **Therneau et al. (1990)** proposed treating the null Martingale residual from the Cox model (1972) as the outcome variable. For a Cox model with time-constant covariates where t_i denotes the observation time for subject i and δ_i is the usual censoring indicator the martingale residual is then given by

(3-18) $\quad \hat{M}_i(t) = \delta_i - \hat{\lambda}_0(t_i) e^{\hat{\beta} X_i}$. [134]

The residuals can be interpreted, at each t, as the difference over $[0,t]$ in the observed number of events minus the expected number given the respective model. As usual, the absolute difference of the arithmetic means of the residuals associated with both offspring daughter nodes is computed for all splits. The split that maximizes the difference is selected.

3.3.1.2 Measures of between node separation

Using the distance between Kaplan-Meier curves, we can split a node with an alternative measure. Heuristically, when two daughter nodes are relatively pure, they tend to differ from each other. More precisely, if one split gives rise to two different looking daughter nodes, each of them is likely to be relatively homogeneous.

It is perhaps easier to think about the situation in the analysis of variance tables where the larger the between variance, the smaller the within variation. Finding two different daughter nodes is a means to increase the between variation and ergo to reduce the within variation. The latter implies the homogeneity of the two daughter nodes.

If we take this point of view, we then *select a split that maximizes the difference between the two daughter nodes or equivalently minimizes their similarity*. For example, we may select a split such that $d_1(S_L, S_R)$ is maximized; here S_L and S_R are the Kaplan-Meier curves of the right and left daughter nodes.

---

[134] Radespiel-Tröger et al. (2003a).

The difference in survival between daughter nodes is usually measured by a two-sample statistic. Here, log-rank test statistic is generally used. We already know that this statistic is the commonly applied approach for testing the significance of the difference between survival times of groups.

Motivated by this fact, **Ciampi et al. (1986)** and **Segal (1988)** suggested selecting a split that results in the largest log-rank test statistic.[135] The larger the value of the test statistics the larger the difference in survival distributions of the two subgroups. *The optimal split is finally selected such that the two resulting subgroups have the largest difference in survival profiles.*

Test statistics have been proposed by a number of other authors, too, as a means to determine the best split. Among them Segal and Bloch (1989), LeBlanc and Crowley (1993, 1995), Intrator (1995), and Butler et al. (1991). The general feature of these proposals is very similar to that of Segal (1988), so that I have decided not to describe these methods in detail.

3.3.2 Pruning – Generation of Optimal Subtree Sequence

Let us remember that in the CART^{TM} methodology, the tree resulting from splitting represented an overfit to the data. In consequence, an algorithm for pruning back the branches of the tree to choose the best tree was applied. In principle, we can proceed as with CART^{TM}, if we manage to define an appropriate node cost $R(\tau)$ for survival trees.

Henceforth, the most convenient way would be to simply use the cost complexity pruning and replace the misclassification cost by another notion. In fact, this is would most authors, that will described below, did.

Besides, in the CART^{TM} methodology we used different measures to assess the quality of a node in the splitting and pruning methodology. As for the splitting, CART^{TM} employed a number of impurity functions to determine which split results in the greatest reduction in impurity. In pruning, the cost complexity was based on misclassification costs.

Although the intent in both cases was to assess the quality of a node, we used different measures for it. In the survival tree methodology this changes. Most authors have proposed to use the CART^{TM} cost complexity pruning for survival trees by merely replacing the notion of node quality with the measure (or a function of it) they had previously used for splitting.

To recall, we defined cost-complexity as

$R_\alpha(T) = R(T) + \alpha |\tilde{T}|$, where the classification performance of the tree was measured by the sum of the quality of its terminal nodes[136]

$$R(T) = \sum_{\tau \in \tilde{T}} R(\tau).$$

In that way, **Davis and Anderson (1989)** use the negative likelihood deviance $-l(\tau)$ (3-9), **Therneau et al. (1990)** the sum of squares of the martingale residuals (3-18) and **Gordon and Olshen (1985)** the impurity measure $i(\tau)$ (3-7) as a proxy of node quality.

LeBlanc and Crowley (1992), too, use CART^{TM} cost complexity pruning. Logically, here, the node quality is replaced by the full likelihood deviance residual from (3-10).

When saying that one could theoretically mix splitting & pruning proposals of the various authors, I was mainly referring to those that allow for easy inheritance of the CART^{TM} methodology.[137]

LeBlanc and Crowley (1993) introduced the notion of *split complexity* as a substitute for cost complexity in pruning a survival tree.

$LR(\tau)$ is defined as the value of the splitting statistic at node τ. They then define *goodness of split complexity* as

(3-19) $\quad LR_\alpha(T) = \sum_{\tau \in \tilde{T}} LR(\tau) - \alpha(|\tilde{T}| - 1).$

---

[135] These approaches are sometimes referred to as "direct approaches", see Keles, Segal (2002).
[136] We took misclassification costs to assess the quality of a node.
[137] In this way, one might, for instance, use the Davis, Anderson (1989) criterion for splitting and the Gordon, Olshen (1985) criterion for pruning.

Note that the summation above is over the set of internal (non-terminal) nodes and $|\tilde{T}|-1$ is the number of internal nodes. Thus, the penalty for the complexity is taken as the number of splits in the tree. They show that an optimally pruned subtree for any α can be found by a weakest link pruning algorithm, which is analogous to the cost complexity pruning algorithm in CART™.

The negative sign in front of α is a reflection of the fact that $LR_\alpha(T)$ is to be maximised, whereas the cost complexity $R_\alpha(T)$ is minimized. They recommend choosing α between 2 and 4 if the logrank test is expressed in the χ_1^2 form.[138]

3.3.3 Final Tree Selection

In survival tree literature, cross-validation known from the CART™ methodology is drawn on by most authors to select the final tree from the sequence of nested optimal subtrees generated by the adoption of cost-complexity pruning. Cross-validation techniques are in a class of techniques referred to as "resampling methods", which is a set of general purpose methods for evaluating performance of data models by purely data driven means. Generally, cross-validatory estimates (i.e. expected prediction error) are calculated and eventually used to determine the final tree.

Accordingly, **LeBlanc and Crowley (1992)** calculate an estimate of expected one-step deviance for each of the pruned subtrees by V-fold cross-validation and the tree that minimizes the estimated deviance is chosen as the final tree. **Davis and Anderson (1989)** use V-fold cross validation as well. And **Gordon and Olshen (1985)** use X-fold cross-validation.

Even if **Therneau et al. (1990)** did not use cross-validation when proposing Martingale residuals as a means to splitting and pruning the tree, cross-validation can be used herein as well to determine the final tree.[139]

It should be clear, by now, that once the splitting measure allows for cost complexity pruning, final tree selection on the basis cross-validation is an intuitive choice.[140]

Another resampling method is boot-strapping which is used by **LeBlanc and Crowley (1993)**. The boot-strap technique is employed to deflate the value of LR and

---

[138] Singer, Zhang (1999), p. 100.
[139] Hess et al. (1999).
[140] Breiman et al. (1984), pp. 311-313.

select the final tree. I already mentioned boot-strapping in the CART™ methodology. B*oot-strapping is a data-based simulation method for statistical inference* which can be used to study the variability of estimated characteristics of the probability distribution of a set of observations and provide confidence intervals for parameters in situations where these are difficult or impossible to derive in the usual way. The basic idea of the procedure involves sampling with replacement to produce random samples of size n from the original data x_1, x_2, x_n each of these is known as a boot-strap sample and each provides an estimate of the parameter of interest. Repeating the process a large number of times provides the required information on the variability of the estimator and the confidence interval.

3.3.4 Alternative Approaches

The proposals that I will subsequently subsume under alternative approaches are those that could be divided into the established 3-step methodology.

Since the log-rank statistics do not provide a global measure of performance,[141] neither minimal cost complexity nor cross-validation (or boot-strapping) is possible.[142] **Segal (1988)** suggests an ad-hoc bottom-up procedure to determine the final tree after splitting is performed.

The idea is as follows: From the bottom, step up the tree, assigning to each internal node the maximum split statistic contained in the subtree of which the node under consideration is the root. Next, collect all these maxima and place them in increasing order. The first pruned tree of the sequence corresponds to locating the highest node in the tree possessing the smallest maximum and removing all its descendents. The second tree of the sequence is then obtained by reapplying this process to the first tree and so on until all that remains is the root node. Essentially, each internal node is linked with the maximum split statistic contained in the subtree for which the node is the root.[143]

Selecting the final tree from the sequence provided can be done by plotting maximal subtree split statistics against tree size and picking the tree corresponding to the characteristic kink in the curve.[144]

---

[141] It is impossible to decompose the log-rank statistic into between node components.
[142] Split-complexity pruning was developed later, LeBlanc, Crowley (1993).
[143] See Segal (1988), pp. 43 for an example.
[144] Friedman (1985) describes in detail how to determine this "kink".

Another idea is described by Radepiel-Tröger et al. (2003b). Here, a so-called stopping rule is applied; specifically, tree splitting is stopped when the respective p-value for the node crosses a certain threshold.

Zhang (1995) does not consider pruning or tree selection. Later on, he argues **Zhang (1998)** that pruning and tree selection should be done manually from the saturated tree by an expert of the field under investigation.

3.4 Final Assessment

A number of approaches are conceivable of how to assess the performance of the various proposals that have appeared in survival tree literature. I will focus on some comparative studies that have been conducted to see how the various proposals performed. Nonetheless, it should be said beforehand that no final conclusions can been drawn yet as to what performance measure is best and consequently which of the proposals may be seen as the superior method. In fact, this discussion is just reaching its peak with a great number of publications on the way.[145]

3.4.1 Assessment of the Splitting, Pruning and Tree Selection Proposals

Zhang (1995) uses a simulation experiment to compare various splitting criteria and to apply a performance score measuring the capability of the splitting criteria for discovering data structure. Through a repetitive simulation algorithm, he generates 250 datasets for each of a total of 12 groups that differ in assumed censoring- and survival distribution and sample size.[146] He, next, applies the splitting rules of Davis and Anderson (1989), Gordon and Olshen (1985), Segal (1988), and his own. For his splitting proposal, he employs the two normalizations (adaptive and global)[147] and the respective weights[148] in $i_y(t)$ (3-15). The performance of a splitting rule is ranked according to the similarity between the 250 recovered trees and the latent tree.[149]

Overall, the adaptive normalization rule does best in recovering the latent tree structure. The order of performance for the proposals of the various authors is:

---

[145] Email communication with Hothorn and Radespiel-Tröger.
[146] Survival distributions: Exponential, Weibull, Mixture Exponential; Sample size: 250, 500; Censoring distributions: uniform in (2, 4), (0.25, 1.25).
[147] Remember from section 3.3.1.: Adaptive normalization: sum in the denominator of $i_y(t)$ over subjects in node t. Global normalization: sum over entire sample.
[148] Weights were 1:2 for adaptive normalization and 1:1 for global normalization.
[149] For a detailed description of the algorithm applied see Zhang (1995), p. 310.

Adaptive normalization, global normalization, (Zhang, 1995), Davis and Anderson (1989), Gordon and Olshen (1985), and finally, Segal (1988).

In another simulation experiment conducted by LeBlanc and Crowley (1992) their respective method shows similar performances as that of Davis and Anderson (1989) method. Surprisingly, this is also the case for exponential survival times when the latter should be expected to perform better.

Crowley et al. (1995) investigate the Gordon and Olshen (1985) splitting proposal. They point out that the distances used as impurity criterion by them tend to find structure in the censoring variable when there is no structure in the underlying covariate space. Conversely, they state that except for the Gordon and Olshen (1985) approach, the current methods are not devoid of parametric assumptions about the underlying hazard model.

Crowley et al. (1996) suspect that the split complexity might be subject to bias. They suggested a bias-corrected version of the split complexity.[150]

Keles and Segal (2002) examine the results of Martingale residuals and log-rank statistics as splitting criteria. Albeit results were found to be similar, they conclude in recommending Martingale residuals for their easy inheritance of the CART™ algorithm with attendant cross-validation.

What has become clear, yet, is that pruning based on between-node separation is conceptually harder as it does not allow for easy inheritance of the CART™ methodology.

Conclusively, it should be said that the paucity of comparative studies and their differing results do not allow for a final assessment of the existing proposals. Besides, there are only few proposals for single performance measures that have come up so far. This is, of course, not surprising given the novelty of the method.

In the classification and regression framework, the goodness of prediction is measured by the misclassification error or mean squared error[151,151]. However, there is no obvious goodness of prediction criterion for predicted survival probability functions. Several proposals have been studied.[152,153]. Apart from simulations, the statistical methodology to assess the adequacy of a model consists mainly of a range of ad-hoc

---

[150] Crowley et al. (1996), p. 21.
[151] Hothorn et al. (2004), p. 79.
[152] Korn, Simon (1990), Altman, Royston (2000).

approaches, and there is a consistent lack of commonly accepted standards in this field.[154]

One ad-hoc approach that is frequently used in survival analysis is the assessment of the Kaplan-Meier estimates of the survival probabilities in the subgroups accompanied by p-values of the **log-rank test** to check for homogeneity across risk strata. I have already described this method in section 2.4.1.

It will come as no surprise that I have decided to use this approach for assessment of the method's goodness of prediction being aware of the fact that it does merely provide a first proxy of it.

A number of other approaches are based on the **ROC methodology**.[155]

Sauerbrei et al. (1999) state that to assess the value of a given prognostic classification scheme, one should compare the estimated event probabilities with the observed individual outcomes. They propose the **Brier score**[156] which had originally been developed for judging the accuracy of probabilistic weather forecasts. Although this approach has met widespread acceptance,[157] censoring is still a problem for the method and could not yet be solved satisfactorily.[158]

It is of general importance to recognize that the time to an event cannot be predicted adequately. Consequently, in statistics, our aim is to best estimate the probability that the event of interest will not occur until $t*$ at $t = 0$ given the available covariate information for a particular subject.

---

[153] See Henderson (1995) for an overview.
[154] Sauerbrei et al. (1999).
[155] Henderson (1995) gives an overview.
[156] Brier (1950).
[157] Hothorn (2004), Radespiel-Tröger et al. (2003a).
[158] Hothorn (2004), p. 10.

3.4.2 Merits & Deficiencies of the Survival Tree Method

Undeniably, a discussion about the merits and deficiencies of the survival tree method will greatly resemble that of the CART™ method. Although generally true, it has to be pointed out that *the (relative) contribution of a new method to a certain area will essentially depend on the methods already existent in the area and the characteristics of the area as such.* More precisely, survival trees may provide more insights in the area of ADT due to the process characteristics of the field under investigation and the previously outlined necessity to recognize and model it as such.[159]

One may, additionally, wonder why merits & deficiencies of the methods are assessed before actually applying it. I believe, it would be simply inappropriate to pretend I reinvented the wheel, here.[160] Much more, I want to use the upcoming presentation of a survival tree for technology adoption as a means to confirm, adjust or reject the merits and deficiencies within the relative pattern described above.

Survival analysis using the proportional hazard modelling framework proposed by Cox (1972) coupled with Kaplan-Meier curves has become the standard procedure to evaluate the impact of certain factors on disease development. In practical situations, nonetheless, problems arise using solely this approach. Within the proportional hazard modelling framework there is no natural way to extract subpopulations of different risks. Besides, when common assumptions such as linearity, additivity and time-constancy of effects are violated, determination and assessment of risk factors becomes unreliable. Thus, alternative approaches are needed to help detect more complicated relationships

With the emergence of survival trees a powerful alternative and supplement for event history data analysis has emerged from survival analysis. While the method provides features that clearly constitute advantages over the Cox model, it also has a number of deficiencies. Each of these merits and deficiencies will be discussed in detail, now.

A key **merit** of the survival tree method is undeniably its easy and straight forward **interpretability**. The final tree provides a decision rule based on information which is easily understood and interpreted regarding the predictive structure of the data.[161] Important inputs are quickly picked out and the number of subsets created is in some

---

[159] See section 2.1-2.3.
[160] Of course, I am not the first assessing this.

sense minimal. The tree-shaped diagram is a powerful way of showing the outcome of analysis. Graphically, it identifies significant predictors and clusters exhibiting significant differences with respect to the dependent variable.

Another great und widely cited advantage is its ability to discover variable interaction that traditional models fail to give. More precisely, it detects **higher-order interdependencies**. As we will see later, by classifying the subjects into subgroups that are determined by $x+1$ ($x = 0,1,2,i$) predictors, higher order dependencies are determined.

A well known problem in regression methods is **multicollinearity**. A crucial condition for the application of least squares, for instance, is that the explanatory variables are not perfectly linearly correlated. The term multicollinearity is used to denote the presence of linear relationships, or near linear relationships, among explanatory variables. If the explanatory variables are perfectly linearly correlated, that is to say, if the correlation coefficient for these variables is equal to unity, the parameters become indeterminate. In consequence, it becomes impossible to obtain numerical values for each parameter separately and the method of least squares breaks down.

At the other extreme, if the explanatory variables are not intercorrelated at all,[162] the variables are orthogonal and there are no problems concerning the estimates of the coefficients, at least as far as multicollinearity is concerned. Actually, in the case of orthogonal Xs there is no need to perform a multiple regression analysis: Each parameter bi, can be estimated by a simple regression of Y on the corresponding regressor: $Y = f(Xi)$. With survival trees these problems cannot occur, because the method circumvents this problem as there is no equation that could possibly collapse because of it.

It is generally acknowledged that data quality is a point of major concern in the construction of a database, and subsequent analyses ranging from simple queries to complex data mining. The quality of knowledge extracted with data mining algorithms is evidently largely determined by the quality of underlying data.

One important aspect of data quality is the proportion of **missing values**. In many applications of data mining a, sometimes, considerable part of the data values is missing. This may occur because the data values were simply never entered into the

---

[161] De Rose, Pallara (1997).
[162] I.e. if the correlation coefficient for these variables is equal to zero.

operational systems from which the mining table was constructed, or because, for example, simple domain checks indicate that entered values are incorrect.[163] Another common cause of missing data is the joining of not entirely matching data sets, which tends to give rise to monotone missing data patterns. Despite the frequent occurrence of missing data, many data mining algorithms handle missing data in a rather ad-hoc way, or simply ignore the problem.[164] Furthermore, the assumptions underlying the way missing data are handled are often not clear, which may lead to erroneous results if the implicit assumptions are violated.[165]

Survival trees handle missing data by so-called surrogate splits[166].[167] Explicitly, in determining whether to send a case with a missing value for the best split left or right, the algorithm uses surrogate splits. It calculates to what extend alternative splits resemble the best split in terms of the number of cases that they send the same way. This resemblance is calculated on the case with both the best split and alternative split observed. Any observation with a missing value for the best split is subsequently classified using the first, most resembling, surrogate split, or if that value is missing also, the second surrogate split, and so on.

Singer and Zhang (1999) stress that if surrogate splits are used, the user should take full advantage of them. In particular, a thorough examination of the best surrogate splits may reveal other important predictors that are absent from the final tree structure, and it may also provide alternative tree structures that in principle can have lower misclassification costs than the final tree, because the final tree is selected in a stepwise manner and is not necessarily a local optimizer in any sense.[168]

Even if frequently quoted as such, one cannot unanimously say that the mere fact that a method is **non-parametric** is an advantage. Admittedly, the absence of parametric assumptions can be an advantage depending on the amount of information available about the process under investigation; once a parametric assumption matches the process, however, the resulting model is difficult to beat.[169]

---

[163] Feelders (2001).
[164] A missing value in any of the covariates results in an undefined linear combination, see Zhang, Bracken (1996) for more details.
[165] Feelders et al. (1998).
[166] Breiman et al. (1984).
[167] There is another, less popular, approach in survival tree literature to deal with missing data. Zhang et al. (1996) named this other approach "missing together" (MT) approach which was first implemented by Clark, Pregibon (1992). I will not introduce this approach in more detail, please see the respective literature for more details. Besides, the idea is not implemented in available software packages. The method of surrogate splits clearly dominates in literature. Nevertheless, the missing together approach offers an interesting alternative
[168] Singer, Zhang (1999), p. 55.
[169] Cox (1995), p. 8.

Let us now turn to **tree stability** and the **variable selection bias,** two related issues widely discussed as being the method's mayor **deficiencies:**

Breiman et al. (1984) call a tree unstable if small perturbations in the learning sample L induce a large change in the predictor. A stable predictor converges to some fixed value as the sample size N tends to infinity, whereas an unstable predictor does not. The stability of a predicted survival probability function derived from a survival tree may be affected by small learning samples, a large number of covariates or a small effect to noise ratio.

As tree instability has been of concern ever since trees appeared in literature, one has continuously looked for remedies:

"Boot-strap aggregation" of classification and regression trees ("bagging"[170]) has been proposed to stabilize predictors in many applications.[171] The idea behind this approach is that the aggregation of multiple unstable predictors leads to a stabilization in many classification and regression problems.

Using boot-strapping, Hothorn et al (2004) propose to derive the predicted survival probability function for a sample member as follows: first a set of survival trees based on B bootstrap sample of the observations is constructed, and second the bootstrap aggregated Kaplan-Meier curve of a new sample member is computed for all bootstrap observations identified by the B leaves containing the new observation. The Brier score confirms improved prediction for this approach.

"Boosting" is another method that has come up to improve tree stability. Boosting was originally proposed by Schapire (1999). It is an iterative re-weighting procedure. At each iteration, observations are weighted by its residual from the previous iteration. The user decides how long to keep at it. At the end, the prediction is the weighted average of all the predictions.[172]

Recently, Shih and Tsai (2004) have shown that selection bias is especially strong for the log-rank approach when predictor variables contain different numbers of missing values.

Lausen et al. (1994) suggest a methodology for trees that accounts for variables measured on different scales. The adjusted *p*-value of a maximally selected log-rank

---

[170] Breiman (1996), Bagging stands for "bootstrap aggregates" see www.stat.Berkley.EDU/users/breiman/.
[171] Sauerbrei (1998) provides a recent overview on boot-strapping in survival analysis.
[172] Q&A Forum S-Plus. Subject: Summary: Bagging & Boosting, website: www.s-news@wubios.wustl.edu.

test is used to adjust for the variable selection bias due to different scales and has shown to improve tree stabilization. Radespiel-Tröger et al. (2003b) showed that this method performed better as the usual maximum log-rank statistic approach[173].

Singer and Zhang (1999) argue that model stability is a generally important issue and deserves serious attention. Yet, it would be ironic to question the tree stability while not being concerned with the model instability in general. Despite the general model instability, the tree structure is not as shaky as it looks. To them the real cause of concern regarding tree stability is the psychological appearance of the tree.[174]

After outlining merits and deficiencies of the method as described in literature, I will now apply survival trees to the area of ADT.

---

[173] Segal (1988).
[174] Singer, Zhang (199), p. 56.

4 The use of Survival Trees to Forecast Innovation Diffusion

I believe it was essential to give an extensive introduction into survival analysis and simultaneously define censored event data in the context of ADT before turning to the inner topic of this thesis. With the survival tree method I will now introduce a new model into ADT, with the reader already having gained the necessary understanding of the general implications of modelling, investigating and interpreting censored event data in the context of ADT.

Let me start by summing up what we already know about the method and what sort of additional insight one can expect method will provide us with.

We know that the method's greatest merits lie in its easy interpretability and his ability to identify higher-order interdependencies that the classical methods, which were described in section 2.4, fail to give. We will now see whether and how these features show up in the context of ADT.

As for the method's mayor deficiencies, there will be only limited possibilities to investigate tree stability and variable selection bias. These problems are well-known to most statistical modelling approaches and thus will be of continuous concern not only in the application of survival trees but in other approaches, too. Moreover, the method's usefulness in the forecasting of innovation diffusion will be determined much more by it merits than by its deficiencies in an area where models ought to be used in a supplementary rather than a competitive manner; hereby providing maximum insight about the diffusion process under investigation.

Consequently, I will concentrate on tree interpretability and identification of higher-order dependencies, in particular, to examine the use of survival trees to forecast innovation diffusion.

It's important to keep in mind how the survival tree method works: subgroups or clusters are identified that are most homogeneous in terms of the risk of the event under investigation. The fundamental difference between survival analysis and ADT lies in the interpretation of an event. This event is death in survival analysis and adoption of an innovation in ADT. Henceforth, the method sorts the observations into clusters that have similar adoption prognosis. That is to say, it makes adoption risk in one node as high as possible and as low as possible in the other node.

Of course, one first has to look for an appropriate software package that allows for survival tree construction. Subsequently, I will apply the method to a dataset coming from a survey taken about the development of e-business adoption in the European Union. The results will be presented and discussed with respect to the central question of this thesis.

4.1 Applicability of Available Software

There are various statistical data analysis and data management packages that may be used to calculate survival trees.

S-PLUS is an object-orientated, command driven programming language that is specifically designed for statistical analysis. Many academic statisticians are using S-PLUS because it allows them the flexibility to modify and combine existing procedures and program new, recently-released methodologies. S-PLUS has been used by a number of authors as working platform among them LeBlanc and Crowley (1992, 1993). Generally, authors do not publish their exact calculations or programming syntax which made it impossible to access and screen them.

The so-called Rpart routine has been developed by Prof. Brian Ripley[175] from the Department of Statistics at the University of Oxford together with Prof. Terry Therneau[176] and Prof. Beth Atkinson[177] both from the Mayo clinic college of medicine. Rpart can be used not only for classification and regression trees but for survival trees as well. For this, a number of modifications have to be taken, so that Rpart recognizes that the variables are survival objects. The inherent splitting criterion corresponds to the one used by LeBlanc and Crowley (1992).[178] As the node quality criterion allows for inheritance of the CART^{TM} methodology, cost complexity pruning can be used with attendant cross-validation.

R is an open-source software that is tailored after the original S-PLUS system. It is not a menu-driven system but posses all features also available in S-PLUS. And most importantly, it is free.[179] Andrea Peters[180] and Torsten Hothorn[181] have developed the so-called Ipred[182] package in R that is based on the Rpart routine and

---

[175] Ripley@stats.ox.ac.uk.
[176] therneau@mayo.edu.
[177] atkinson@mayo.edu.
[178] Technical Report, 61, p. 36, 41, www. mayo.edu/hsr/biostat.html.#61.
Email correspondence Terry Therneau, Beth Atkinson.
[179] It can be accessed through www.r-project.org.
[180] Andrea.Peters@imbe.imed.uni-erlangen.de.
[181] Torsten.Hothorn@rzmail.uni-erlangen.de.
[182] http://cran.r-project.org/doc/packages/ipred.pdf.

applies the bagging procedure proposed by Hothorn et al. (2004). R has the same relatively long learning curve as S-PLUS.

Rtree has been developed by Professor Heping Zhang[183] at Zhang's Lab of statistics & Biometrics, University of Yale.[184] It implements the ideas of Zhang (1995). The software allows for choosing each split separately. Nevertheless, the software does not permit practical integration of external data and exhibits other features in final data and graphical output that excluded the possibility of using it.[185]

Xplore is a statistical software that has been developed at Prof. Haerdle's Institute for Statistics & Econometrics, Humboldt University, Berlin. Xplore offers a module called STREE[186] that is quoted to implement the ideas of Singer and Zhang (1999), and Zhang (1995). Unfortunately, the free ware version does not include this module and allows only for a maximum of 500 observations.[187]

TSSA is a software package that interfaces with S-Plus and R.[188] It implements the ideas of Segal (1988). Prof. Segal pointed out that the version was old and that compatibility with newer operating systems has not been checked yet. So it came as no surprise that the program didn't work under XP or Windows 2000.[189]

Although **SAS** is the data analysis package used by most applied statisticians, I didn't find evidence on any macro implemented in SAS that would allow for survival tree calculation.[190] The package is commercially distributed and would have had to be excluded for cost reasons in any case.

As mentioned, a great number of authors have developed their own applications. Usually these applications are neither user-friendly nor publicly accessible which, of course, doesn't imply anything about the proposal as such. An exception, nonetheless, is Radespiel-Tröger who has vowed to publish his application, which is most likely to be customized for external use, soon.[191] Additionally, I was looking for software that could improve graphical output of the various software packages.

---

[183] heping.zhang@yale.edu.
[184] www.masal.med.yale.edu/rtree.
[185] Many helpful comments on how to use this software came from Prof. Zhang's assistant, Yuanging Ye, yye@masal.med.yale.edu .
[186] The software can be assessed through www.xplore-stat.de.
The TSSA module has been developed by P. Cizek, K. Komorad, W.Haerdle (2002).
[187] Special thanks to Uwe Ziegenhagen, mdtech@mdtech.de.
[188] www.stats.ox.ac.uk/pub/SWin/tssa.zip.
[189] Email contact Prof. Segal, mark@biostat.ucsf.edu.
[190] You can see more about the package by linking to www.sas.com.
[191] Email exchange Radepiel-Tröger.

Klimt[192] is a software package that has been developed by Simon Urbanek[193] at the University of Augsburg's Department of Computer Orientated Statistics to allow for interactive graphical representation and pruning of a tree.[194] Klimt has won the "chambers software award" in 2002. The program is supposed to interface with S-PLUS and R. However, KLIMT is still in work progress and is not supported by Windows.[195]

4.2 Data Description & Handling

The data that will be used to forecast innovation diffusion on the basis of survival trees comes from the e-business W@tch which has been set up by the European Commission in 2001 to monitor "the adoption, development and impact of electronic business practices in different sectors of the European economy".[196] The data comes from their newest available survey (2nd survey Nov. 2003) that includes a total of 7.302 observations. Observations were recorded in 25 countries which consisted mainly of EU member states. Decision-makers in 11 sectors were interviewed about their company's e-business adoption history with sector coverage differing among countries.

Bias in the data coming from heterogeneous sector coverage (Appendix 6.2) was of indisputable concern. Initially, I considered to exclude countries from the analysis where less than three sectors were covered or where the total number of observation was less than 100 but eventually came to the conclusion that the heterogeneity in the data could not be reduced by this and that the issue would have to be watched closely anyway.

Among the variables recorded I included only those that were considered as being adequate predictors of an e-purchase adoption distribution. The final number of predictors was 52. Appendix 6.3 gives an overview of the variables and documents final data handling.

A number of variables could be identified as control variables giving me the possibility to exclude observations from the analysis where the pre-requisites of e-business adoption were not fulfilled in the first instance (e.g. a company had no internet

---

[192] **K**lassification – **I**nteractive **M**ethods for **T**rees.
[193] Simon.Urbanek@math.uni-augsburg.de.
[194] http://stats.math.uni-augsburg.de/Klimt/.
[195] Email correspondence Simon Urbanek.
[196] All publications of the e-business w@tch including the statistical database, booklet, and various other related

access). I set the beginning of our observation period at 1993 and excluded all observations where a company stated that they had adopted before this year for implausibility reasons.[197] This reduced the final number of observations to 5.615.

Furthermore, most variables were recorded as dummies to enable easier subgroup-specific splitting differentiation.

Adoption time was recorded at yearly basis. Although one might question whether one could really talk of T being measured continuously here, the data was used as such. I would like to stress that the essential point remains to see whether the survival trees can be used to the benefit of ADT. Therefore, I will not put to much attention to this issue. It will be on later investigations to improve data quality, once the method has shown to be of any benefit. Variables were all time-invariant.

The time-to-adoption was measured for e-sale, e-purchase, supply chain management, customer relationship management, knowledge management, and enterprise resource planning. [198]

Table 4-1 gives a first overview of the adoption distribution of the respective e-business technologies. E-purchase is the e-business innovation that is earliest adopted (by average *8.29* years).[199]

When calculating the average adoption time, we have to handle the 2002* data that is censored as if firms had adopted after nine years, so that within the calculation, we assumed that 3.989 firms would have adopted, although this certainly is not the fact. Nevertheless, the average adoption time provides us with a first quick idea of the process under investigation, although the problem of censoring cannot be handled convincingly.

---

information is available in electronic format under www.ebusiness-watch.org .
[197] This point in time had to be set somehow arbitrarily and may be questioned. In early 1993, however, the first graphical user interface, mosaic for X, emerged, making the internet for the first time "readable" for the public.
[198] Check the website for a definition of the various e-business technologies.
[199] Remember that the start of the observation period is 1993.

Table 4-1: E-business adoption pattern for recorded dependent variables

	Esale	Epurch	SCM	CRM	KM	ERP	Elear
Adoptions	954	2585	217	626	368	645	536
1993	0	0	0	0	0	0	0
1994	5	7	2	7	7	20	2
1995	16	33	1	11	8	23	3
1996	17	41	6	11	17	31	5
1997	41	81	9	24	12	43	14
1998	71	199	13	41	25	43	20
1999	114	260	15	37	19	72	23
2000	164	477	30	86	51	79	71
2001	206	528	21	97	48	68	88
2002	320	959	120	312	181	266	310
2002*	4661	3030	5398	4989	5247	4970	5079
% Censored Data	83,0%	54,0%	96,1%	88,9%	93,4%	88,5%	90,5%
Mean Adoption Time	8,71	8,29	8,95	8,85	8,90	8,76	8,91

I decided to investigate the process of e-purchase adoption as the means to see whether survival trees could be used for ADT.

The respective question in the questionnaire was: "When did your company purchase goods or services online for the first time."[200]

The process of e-purchase adoption is depicted in figure 3-4 now using the 1-survivor function that I trust to give a more intuitive graphical estimation of the process in the given context.[201]

Figure 4-1: 1-Survivor function for e-purchase adoption

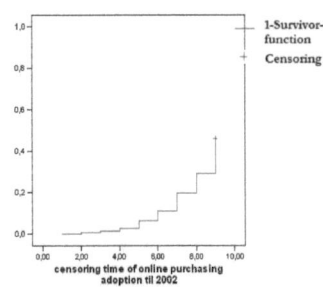

In this form, the Kaplan-Meier representation of the survivor function reads as follows: the probability that a company has already adopted increases each year. In 2002 the probability of having already adopted e-purchase lies at 56% implying that 54% of the data is censored. E-purchase is the process with the highest number of

---

[200] www.ebusiness-watch.org/menu/The_European_e-Business_Survey/ , p. 15.
[201] See section 2.4.1.

adoptions (2,585) within the observation period. In 2000, the large rise in the curve is caused by the relatively large number of adoptions in that year (959).

To gain more than a quick overview of the data structure let us now turn to survival trees for multivariate analysis.

4.3 Implementation

I decided to use R to calculate the survival tree for e-purchase adoption.[202] Other options had to be excluded for cost or inapplicability reasons.[203] It has to be remarked, nevertheless, that the choice taken offered a convincing solution, as it used S-PLUS features and the proposal, on which the application was developed, could be identified precisely.

Moreover, log-rank tests were calculated for some of the terminal nodes to give some form of a check of the goodness of prediction for the method and its resulting output. Although the log-rank statistics can be calculated in R as well,[204] I decided to calculate them with SPSS[205] for higher calculatory and graphical convenience.

As remarked, I used the Rpart package in R which was also used to calculate the classification and regression tree in section 3.1. The data had to be modified so that Rpart recognized the dependent variable as survival object and applied the LeBlanc and Crowley (1992) proposal, accordingly. Appendix 6.4 displays the syntax used for tree calculation. I have added short explanations so that each step can be identified independently.[206]

The final tree is determined on the basis of the 0-SE rule. Here, that final tree is chosen that minimizes the expected prediction error that is determined through cross-validation.[207] Appendix 6.5 displays the respective data output from R. One can see that through the respective complexity parameter (cp), the tree is determined (i.e. the tree is trimmed to that complexity parameter). The minimum number of observation was set to 20 just as LeBlanc and Crowley (1992) had done.[208]

I have summarized the LeBlanc and Crowley proposal in table 4-2.

---

[202] Newest available version was: R 1.9.1.
[203] See section 4.1.
[204] The respective function is survfit().
[205] Trial version can be downloaded under www.spss.com.
[206] The syntax can be simply copied into R and used after respective data-location modifications.
[207] Another possibility is the 1-SE rule. Here the standard error is added to the average error and the first tree whose average error lies beyond this value is chosen.
[208] Extremely small prognostic groups are not of interest for staging.

Table 4-2: Summary of LeBlanc and Crowley (1992) proposal

Splitting Criteria Based on (full likelihood) deviance	Pruning Analogous cost-complexity pruning	Final Tree Selection
$R(\tau) = \frac{1}{N} \sum_{i \in S_T} \left[\delta \cdot \log\left(\frac{\delta}{\hat{\Lambda}_0(ti)\hat{\theta}\tau}\right) - (\delta - \hat{\Lambda}_0(ti)\hat{\theta}\tau) \right]$	$R_\alpha(T) = R(T) + \alpha\|\tilde{\tau}\|$	X-fold cross-validation

4.4 Results

I re-built the final tree from R for higher convenience in graphical output using the accompanying numerical output. Inevitably, one puts higher demands on the graphical output of a method whose apparent strength lies in easy interpretability. Appendix 6.6, 6.7 show the saturated and the final tree originally generated by R.

Nonetheless, there are a number of options given in R to improve the graphical output of the tree. Unfortunately, the larger the tree becomes the worse tree visibility becomes, too.[209]

The final survival tree for e-purchase adoption consists of 24 internal and 25 terminal nodes. The tree reads as follows: The nodes depicted contain information about the node number, the mean adoption time and the splitting criteria on which the respective node was split into two. In this way, the root node (1), whose overall mean adoption time is 8.29 years, uses ICT services as the splitting criteria implying that this is the predictor with the highest predictive power for the initial dataset.

The node numbers indicate the timely order of the splitting process. Through this, the nodes/branches that have been pruned off can be identified, as well

The last node that has survived the pruning process is number 98,244 implying that 98,243 splits had been processed before arriving at this node.

Observations that go to the right exhibit relatively earlier mean adoption times whereas those that go to the left contain observations with relatively late mean adoption times. One can see that splitting observations by their ICT service sector affiliation reduces the mean adoption time by approximately one year.

---

[209] I applied all available options but did not arrive at a tree that I considered convenient enough.

The subgroup adopting e-purchase earlier than any other of the depicted one is to be found in terminal node 13 (6.78 years). This node contains Czech companies from the sector of ICT services.

Let us compare this node with terminal node number 20. This node contains companies belonging to any other than the ICT service sector from Greece whose group size (i.e. number of subsidiaries of the company) is lower than 9. These companies are expected to adopt e-purchase after 8.97 years, that is, more than 2 years later than the Czech ICT service companies.

Figure 4-2: Survival tree for e-purchase adoption

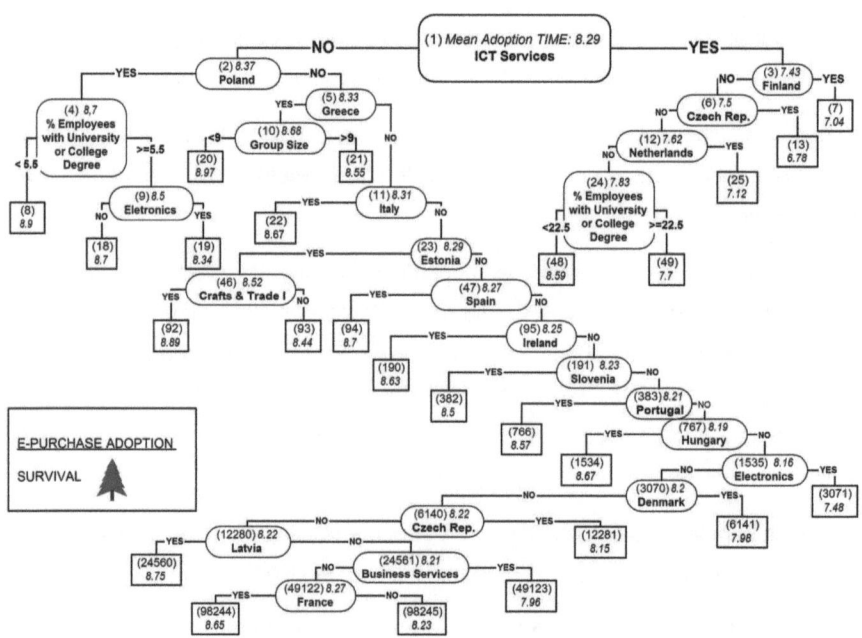

The split that produced the terminal node 20 was based on a numerical value. The implicit question was whether the number of subsidiaries of the respective company was less than 9. Although this particular split was based on a (binary) dummy variable (g01q_1, Appendix 6.3) the method as such is capable of choosing a splitting variable by itself. This unique characteristic of the method can be observed in node number 24 and 4 (g10, Appendix, 6.3). Observations are split according to the percentage of employees having a college or university degree with the method determining the precise percentage by itself.

We clearly, see that country affiliation plays an important rule as a predictor of e-purchase adoption risk. Country of origin is used in 16 out of the 24 splits, sector affiliation in 5, employee's level of education in 2, and group size in 1 case. It is interesting to see that a company's market share (Appendix 6.3., $g11\_1$ to $g11\_5$) and the question of who are the company's primary customers (Appendix, $g06\_1$ to $g06\_4$) are not used as splitting criteria, at all.

Nonetheless, this should not bring us to the conclusion that these co-variables are no good in predicting e-purchase adoption. This has to do much more with the inherent methodology of the survival tree method than with the predictive power of the various variables.

Survival trees have an essentially different way of assessing variable impact on the process under investigation. In fact, the impact of a predictor is assessed on each dataset only once. In other words, the second best predictor of the whole dataset may never show up in a survival tree as the pattern of the dataset changes with each split.

In the classical event history models in section 2.4, variable impact is ascertained for each variable on the whole dataset. In this way, one gets the impact of all co-variables for this one dataset and can subsequently rank them according to their predictive power.

In survival tree methodology there is only one best predictor for one specific dataset that changes as the method proceeds. Although R displays the second and the third best predictor, as well, only the predictor offering the highest reduction in impurity is chosen and does eventually show up in the final tree. Surrogate splits offer additional insight but should not be confused with second or third best predictors as they are calculated again on another dataset; one where missing variables were excluded.

Ultimately, in the survival tree method one gets the best predictors for a great number of subsets of the data. This way, subgroups are detected that exhibit significant differences in adoption risk. This is undoubtedly something entirely new and deserves special emphasize. The insight that this method offers is thus substantially different to that we would get from a parametric or the semi-parametric model. Anything like a competitive comparison would be out of place here.

The data output for node 4 is displayed in table 4-3. A number of information is contained such as the estimated rate which is calculated as the number of events over total observation time, scaled so that overall event rate is 1 for the root node.[210] The mean deviance indicated must correspond to the estimated one-step deviance from 3-13. Remember, that the tree that minimizes the mean deviance/expected prediction error, calculated via cross-validation, is chosen as the final tree.[211]

Table 4-3: Data output for final tree

```
Node number 4: 489 observations,   complexity param=0.002223256
  events=115,  estimated rate=0.4559475 , mean deviance=0.961087
  left son=8 (112 obs) right son=9 (377 obs)
  Primary splits:
      g10   < 5.5 to the left, improve=17.519000, (47 missing)
      z01b.3 < 0.5 to the left, improve=14.626860, (0 missing)
      g11.4 < 0.5 to the left, improve= 5.332081, (0 missing)
  Surrogate splits:
      z01b.1 < 0.5 to the right, agree=0.776, adj=0.057
```

Information on a surrogate split is depicted, as well, because the $g10$ co-variable (% of employees with a college or university degree in company) contains 47 missing values for those observations where companies said they couldn't answer the question ($g10\_dk$).

I conducted a log-rank test comparing the survival curves of the first (node 2, node 3) and the last nodes (node 98244, node 98245) that resulted from the splitting process with each other. Detailed numerical and graphical information is shown in figure 4-3 and 4-4. Both p values show highly significant differences in the adoption rates of the resulting subgroups ($p < 0.001$).

---

[210] Email exchange Therneau.
[211] O-SE rule.

Figure 4-3: Log-Rank test for node 2 * node 3

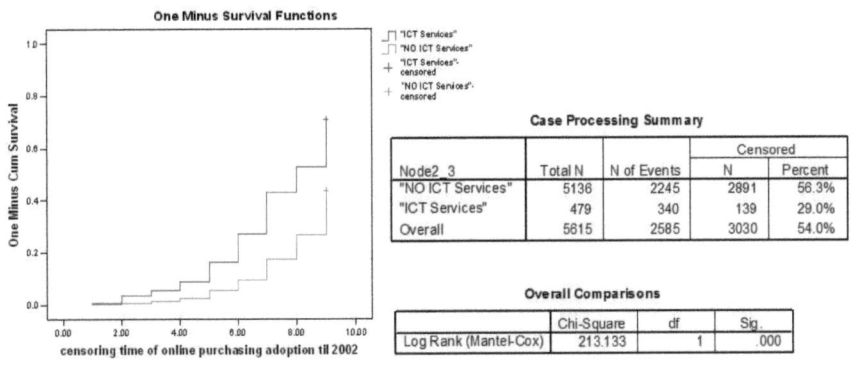

Figure 4-4: Log-Rank test for Node 98244 * Node 98245

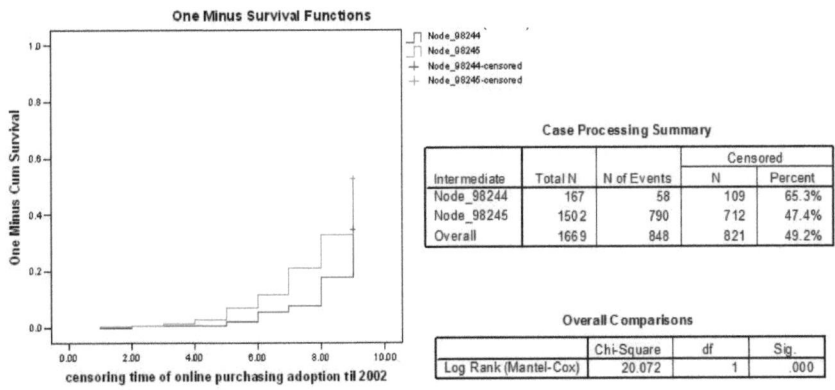

Additionally, I compared the Kaplan-Meier estimate of the adoption curves (i.e. survival curves) of the two most extreme subgroups with the adoption curves of the rest of the dataset in figure 4-5 and 4-6. Again, p values show highly significant differences in the adoption rates of the respective groups.

Figure 4-5: Log-Rank test for Node 13 * Rest

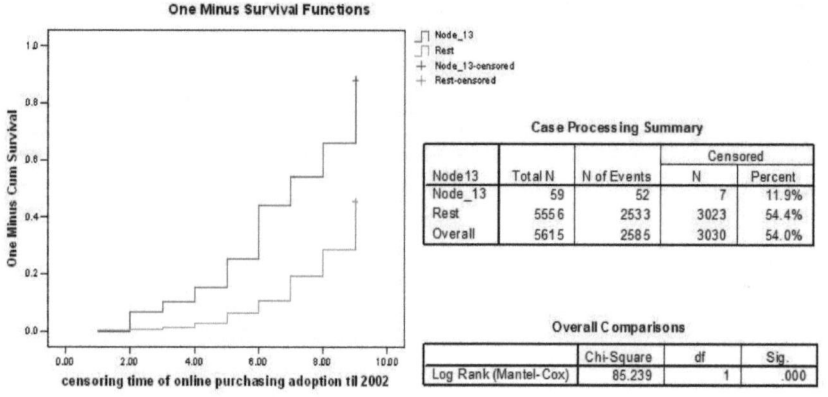

Figure 4-6: Log-rank test Node 20 * Rest

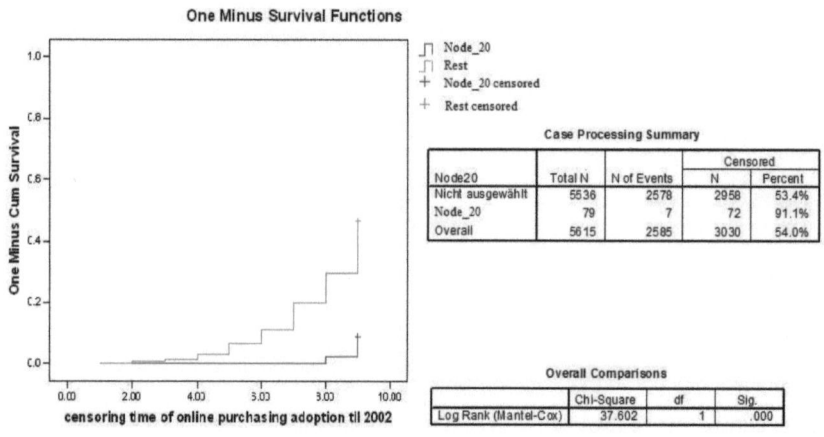

All tests achieve statistical significance proving that survival trees return significant results.

We have seen that the method preserves the merits that were described previously. The insight that this method offers is indisputably new and dooms it to be considered as a vital alternative or supplement to the existing classical event history models.

4.5 Discussion

At the beginning of this thesis, I developed the argument that diffusion models should be dynamic and be based on the micro level. Survival trees comply with both of these demands: Adopter heterogeneity is considered and the process characteristics of the diffusion process are taken into account by means of event history data.

Nonetheless, macro models, although genuinely criticised for neglecting adopter heterogeneity, provide solid answers for questions commonly raised in scientific theory. If we imagine the prognostication of adoption diffusion on the basis of marketing variables, this becomes clear. Therefore, negligence of heterogeneity may be of no great concern in its application. After all, macro models experience continuous popularity and are still more frequently employed than micro models.

Thus, I believe the subject should not be discussed in an exclusive manner making the micro perspective and the dynamic process the necessary condition for a diffusion model. I developed these arguments much more to demonstrate the appropriateness of dynamic micro models for modelling innovation diffusion, as they include vital information that static models and macro models simply deny. Yet, as said, depending on the focus of the analysis, this deficiency may not play the decisive role in applications.

We have seen that survival trees identify subgroups that exhibit significant differences with respect to adoption prognosis. In that way, the method identifies significant groups of early and late adopters. The goodness of prediction has been successfully investigated and although tree stability is a concern, remedies have been developed already and can be expected to expand even more in the future.

Ultimately, the usefulness of this method to forecast innovation diffusion is indisputable and the method can be classified as a new diffusion model.

Let us now turn to the method's usefulness in the adoption- and diffusion context: The identification of significant adopter groups, on the basis of censored event data, is an entirely new insight that the method offers. Especially in an area where the diffusion process is often conditioned by distinction into adopter groups according to the widely accepted Rogers' tradition, this ability promises widespread applicability.

Ironically, the only result of the survival tree method that would offer no additional insight when used together with the Cox model would be the splitting criteria for the root node which, we know, is the one with the highest predictive power.

As the impact of predictors and the adoption process is assessed for the entire dataset and the splitting proposal that we used is based on the ideas of the proportional hazard model, it is no surprise that both models indicate that ICT service affiliation is the greatest adoption accelerator in the data investigated.[212] From here on, insight differs greatly with the survival tree method detecting significant predictors for subsets of the initial dataset.

Furthermore, the ability of the method to distinguish early from late adopters may make it an **interesting tool for innovation marketing** purposes.

In general, the main goal in innovation marketing is to gain product acceptance by the largest number of consumers in the market in the shortest span of time.[213] For this purpose the method could be ideal.

It could help to save resources by targeting merely those customers that are most likely to adopt in the early stage of the diffusion process and who are essential in driving the process to its critical mass. Let us take into account Rogers' widely confirmed theory of early adopters serving as opinion leaders who persuade others to adopt the innovation by providing evaluative information.[214]

In the marketing world there are few jobs as daunting as that of launching a new product. About 80 percent of new products will fail within the first three years of introduction.[215] Hundreds of thousands of dollars are invested yearly to develop, design, and market new products in a saturated environment with short purchase cycles and selective customers.

The success of a new product is often determined early in its introduction. Suboptimal marketing can kill a new product idea before it has a chance to thrive. At the same time the, often inflexible, marketing budget has to be spent wisely along the diffusion process. It is therefore critically important to gain a pulse on how a new product is faring the market place early in the process.

---

[212] Likewise, the appearance of this predictor in the root node of the classification and regression tree in figure 3-1 should not surprise us to much.
[213] Cateora (1993), p. 383.
[214] Rogers (1995).
[215] Wilke, Sorvillo (2004).

For marketing purposes the four adopter categories developed by Rogers have proven to be impractical. Thus, in the marketing world one typically distinguishes between two principal consumer groups in the marketplace: [216] Those who are eager to try a new product with unknowns, early adopters, and those who prefer to wait until others have tried the product first, late adopters. The early adopters provide significant insight into overall consumer acceptance and are the most likely candidates to determine a new product's success. Early adopters tend to communicate their likes and dislikes with others and they adopt new products quickly.

Feedback from early adopters helps marketers address business opportunities and fine-tune strategies early in the process in order to achieve maximum success in the marketplace. It is vital to recognize that new products have the opportunity to be new with the consumer only once, which is why early understanding of consumer adoption is critical in driving business optimisation. Given the leadership that early adopters represent, they often act as a barometer and are critical to a brand's long-term viability and overall success.

So far, many techniques have been applied to select the targets in commercial applications, such as statistical regression[217], regression trees (CART^{TM})[218], neural computing[219] and cluster analysis[220]. None of these method, yet, may match the above outlined innovation marketing framework as good as survival trees by automatically telling a marketer which groups to target and which ones not. Besides, these methods are static.

One issue that, I believe, should be screened more closely is **the way the method takes censoring into account** and what implications that has on the prediction quality. I have already stated that mean adoption time provides merely a rough description of the data distribution in a clustered group. I included information about mean adoption time in the respective nodes and we saw that the method consistently split the data into subgroups with higher and lower mean adoption times.

Adoption time would undoubtedly be the proper number if there were no censored data, but should we really use it for censored data?

To answer this question, I sense, it is important to understand how the survival tree

---

[216] These ideas have been developed linking to: http://www2.acnielsen.com.
[217] Bult (1993).
[218] Haughton, Oulabi (1993).
[219] Zahavi, Levin (1997).

method (i.e. the LeBlanc and Crowley/1992 splitting proposal) takes censoring into account. For this reason, I constructed 4 different groups of hypothetical innovation adopters for 2 different scenarios in figure 4-7.

Let us explore the results from the survival tree imagining that we would be marketers trying to identify whom to target in the early process of the innovation diffusion process.

In **scenario 1**, there is no censored data at all. Data was distributed in a way that groups range from early to late adopters. As before, I have included information about node number, mean adoption time (MAT) and splitting criteria in each node. Additionally, I have incorporated the estimated rate (ER) that I want to take a closer look at. Unchanged, for splits that go to the right, mean adoption time decreases, whereas for those that go to the left, it increases.

Figure 4-7: Scenarios with differing censoring structure

Scenario 1

	Group1	Group2	Group3	Group4
Observations	20	20	20	20
Early Adopters (1)	20	19	10	5
Late Adopters (9)	0	1	10	15
MAT	1	1,4	5	7
Overall	3,6			

Scenario 2

	Group1	Group2	Group3	Group4
Observations	20	20	20	20
Early Adopters (1)	20	19	10	5
Late Adopters (9)	0	1	10	15
MAT	1	1,4	5	7
Overall	3,6			

☐ Censored Data

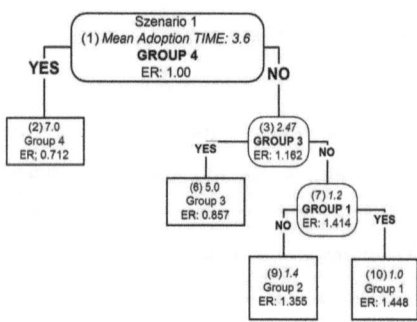
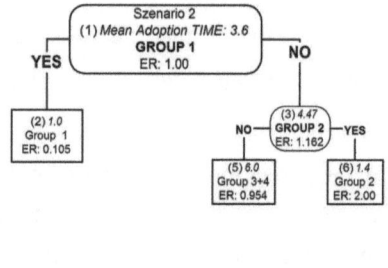

---

[220] Setnes, Kaymak (2001)

In **scenario 2**, only one thing changed with respect to scenario 1: While before no data was censored in any group,[221] now all data in group 1 is censored. Notice that average adoption time has remained unchanged in both scenarios. What will the method now tell us about subgroup targeting?

In scenario 1, the first split is based on group 4 affiliation. As a result, the split with the highest predictive power splits into subgroups containing either group 4 (late adopter) or group 1, 2, and 3 (early adopter). Eventually, we get the group with the highest average adoption time as the node to the extreme left and the group with lowest mean adoption time as the node to the extreme right. The scenario, of course, is of no use for showing the merits of the method, but this is not important now.

Translating our results into the innovation marketing world, the method tells me to target those groups that go the right early in the diffusion process. The first split causes the groups to be arranged into the two subgroups with the highest and lowest possible difference in adoption rates. In other words, if we were asked to divide the market into early and late adopters (i.e. if we had one split only), this would be the best distinction.

Now, under the changed censoring structure for group 1 would it still make sense to tell us to target group 1? Here, all data is censored and although this took place at $t = 1$, it could be fatal to target this group early in the diffusion process.

Eventually, In scenario 2, group 1 is split to the left. Suddenly, the indication of mean adoption time is not consistent anymore with the splitting procedure, with the group that exhibits lowest adoption time going to the left. Consequently, the method tells me to target any other than group 1 early in the diffusion process. Why is that?

If we look at the measured event rate, the indicated rates are conform with the method with higher event rates going to the left and lower ones going to the right.

The calculation of event rate takes censoring into account with every censored data inevitably diminishing event rate. The problem with the event rate is, however, its more difficult interpretability. This is why mean adoption time was initially used as it allows for easier interpretation.

We see, here, that much more needs to be learned and understood about the

---

[221] Interesting to see that static methods, such as logistic regressions or the CART™ method, judge all 4 groups equally with the snapshot taken at the end of the observation period. Henceforth, in all groups adoption is 100%. No conclusions could be drawn with respect to any marketing timing.

method and the interpretation of the data. It should have become clear, however, that the method does NOT identify subgroups with highest or lowest mean adoption time, but with respective event rates or adoption rates, as I called it.

Finally, I want to discuss an issue raised at the very beginning when discussing the inability of the disaggregate models to establish the aggregate level. Here, I referred to methods such as logistic regressions. As for survival trees, by identifying predictive groups of early or late adopters, the question of how to generate the aggregate level does not have to be posed anymore. As a matter of fact, the method is both a micro and a macro model.

I consciously will leave the discussion open. It should be clear that this thesis is merely a start. More research and understanding is necessary to effectively integrate this method into ADT.

5 Summary

At the beginning of the thesis, I developed the argument that dynamic micro models should be used to forecast innovation diffusion. Models from the area of survival analysis, so-called event history or survival models, were found to be uniquely suited for this.

Event history models capture the dynamics of the diffusion process and respect adopter heterogeneity. For this reason, censored event data and its respective methods were embedded into ADT to establish the necessary pattern for the introduction of the survival tree method into the area.

Survival trees, a novel method that has emerged in the area of survival analysis, were subsequently introduced. It was intended to provide a complete overview of the various approaches that have been developed so far. Eventually, the method was applied on a dataset about the adoption of e-purchase by companies coming mostly from EU-countries.

We saw that the method's merits provided new insight into the diffusion process that no other method from the area had been capable of providing so far. The method's easy interpretation will certainly accelerate its acceptance in ADT and will make it accessible to a much greater audience than the other methods. In innovation marketing, especially, the method might prove a powerful new tool responding directly to questions typically asked here.

Nevertheless, more research is needed in this area. Interpretation of data output in the ADT has to be watched closely and software applications have to be improved to commensurate the straight-forward graphical output that the method offers.

6 Appendix

6.1 Classification and Regression Tree for E-purchase Adoption

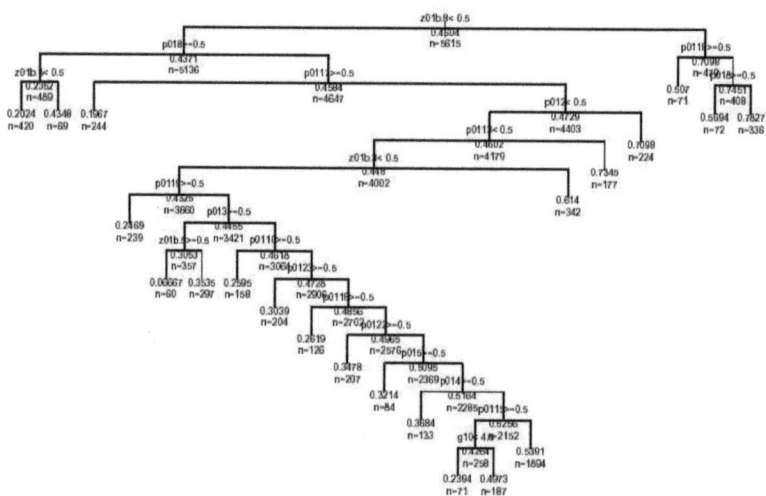

6.2 Cross Table for Sector and Country Coverage

p01: country code * z01b: survey sector no. Crosstabulation

		textile, footwear and leather industries	manufacture of chemicals and chemical products	manufacture of electrical machinery and electronics	manufacture of transport equipment	crafts & trade i	retail	tourism	ict services	business services	health & social services	crafts & trade ii	Total
p01: country code	cyprus	0	0	0	0	0	53	0	0	0	0	0	53
	czech republic	0	55	0	56	0	0	55	59	0	58	0	283
	estonia	44	49	47	20	60	42	46	50	48	48	0	454
	hungary	0	0	66	62	0	0	0	0	71	0	0	199
	latvia	18	37	0	0	0	29	0	0	0	0	0	84
	lithuania	0	0	0	0	0	31	0	0	0	0	0	31
	malta	0	0	0	0	0	0	48	0	0	0	0	48
	poland	43	60	69	56	40	53	61	72	63	44	0	561
	slovakia	34	0	35	0	0	27	0	0	50	0	0	146
	slovenia	0	0	54	0	0	0	50	53	57	51	0	265
	austria	0	0	0	60	0	0	125	0	0	88	0	273
	belgium	0	85	0	0	0	71	0	0	92	0	0	248
	denmark	0	0	0	0	0	55	64	0	0	58	0	177
	finland	53	0	68	0	0	0	0	75	0	0	0	196
	france	68	0	0	0	10	0	0	0	91	59	30	258
	germany	77	0	0	0	0	22	0	0	91	69	58	317
	greece	45	0	56	46	5	0	57	0	0	0	35	244
	ireland	0	67	0	0	0	0	59	71	0	0	0	197
	italy	56	0	0	0	11	0	0	0	0	73	64	35	239
	netherlands	78	0	0	0	0	0	0	99	0	83	0	260
	portugal	0	0	0	63	0	71	0	0	73	0	0	207
	spain	44	0	0	0	0	9	0	0	64	50	37	204
	sweden	0	72	72	70	0	0	0	0	73	0	0	287
	uk	77	0	0	0	17	0	0	0	86	66	48	294
	norway	30	0	0	0	0	60	0	0	0	0	0	90
	Total	667	425	467	433	174	492	565	479	932	738	243	5.615

6.3 Variable Description and Handling

Code	Description	Code	Description
p00	Project no.	z01b	Survey sector no.
p01	Country code	z01b_1	Texuntile, footwear and leather
p011	Cyprus	z01b_2	Chemicals
p012	Czech Republic	z01b_3	Electronics
p013	Estonia	z01b_4	Transport Equipment
p014	Hungary	z01b_5	Crafts & Trade I
p015	Latvia	z01b_6	Retail
p016	Lithuania	z01b_7	Tourism
p017	Malta	z01b_8	ICT services
p018	Poland	z01b_9	Business services
p019	Slovakia	z01b_10	Health and social services
p0110	Slovenia	z01b_11	Crafts & Trade II
p0111	Austria	g03	Number of establishments in (country) - absolute
p0112	Belgium	g03_dk	Number of establishments in (country) - dk
p0113	Denmark	g10	% of employees with a college or university degree in company
p0114	Finland	g10_dk	% of employees with a college or university degree in company - dk
p0115	France	g11	Market share of company in the region you consider your main sales area
p0116	Germany	g11_1	Less than 1% market share
p0117	Greece	g11_2	1% - 5% market share
p0118	Ireland	g11_3	6% to 10% market share
p0119	Italy	g11_4	11% to 25% market share
p0121	Netherlands	g11_5	> 25% market share
p0122	Portugal	g11_6	don't know market share, refused, not applicable
p0123	Spain	g01quote	g01quote: quota group - size
p0124	Sweden	g01q_1	1-9
p0125	UK	g01q_2	10-49
p0126	Norway	g01q_3	50-249
a01	Does company use computers?	g01q_4	250+
a02	Does company have access to the internet?	g02	One or more establishments
a06_a	Does your company use - e-mail?	g02_1	One establishment
a06_b	Does your company use the world wide web?	g02_2	More than one establishment
a06_c	Does your company use - an intranet?	g06	In general, which of the following groups are the primary customers of your company
a06_d	Does your company use - an extranet?	g06_1	consumers
a06_e	Does your company use - a local area network - lan?	g06_2	other businesses
a06_f	Does your company use - a wide area network - wan?	g06_3	public sector
a40_a	Employee training: does your company offer in-house computer or it training	g06_4	no primary customers - mixed
a40_b	Does your company offer participation in IT or computer training offered by third party?		
a40_c	Employees can use some of their working time for learning activities		
a40_d	Does your company offer - any other support measures?	epurch_t	t to online purchasing adoption (observation period ends 2002)
a41	Has your company recruited or tried to recruit staff with special IT skills during the last 12 month?	epurch	e-purchasing censoring indicator
a42	Has your company experienced difficulties in finding staff with special IT skills?		

Excluded as predictors (Reasons: not suitable, bias reduction, described through dummy variables)

Outcome Variables

Predictors

Control variable

6.4 R-Syntax for Survival Tree

```
#CALCULATION OF SURVIVAL TREE FOR E-PURCHASE ADOPTION
#Remove all stored R-objects
rm(list=ls(all=TRUE))
#Read calculation time
Start.Time<-Sys.time()
# Load libraries necessary to conduct calculations in "R"
library (foreign)
library(rpart)
library(survival)
#Read external file stored by Excel in .csv format
Epurchase.df<-read.csv ("C:/Dokumente und
Einstellungen/Burkhard/Desktop/Diplomarbeit/Computational/Data
base/Modified/Ebiz_2nd_s_2003.csv", header=TRUE, sep=";", dec=",",
na.strings="NA")
#Give out Data as table to check for completeness of data
#GROWING THE TREE
#PRUNING - GENERATION OF SEQUENCE OF OPTIMAL SUBTREES
#Define control variables
temp1<-rpart.control(xval=10, cp=0, mibucket=20, minsplit=2, maxcomp=0, )
#Calculate saturated survival tree + generate optimal subtrees
fit1<- rpart(Surv(epurch.t, epurch) ~ p012 + p013 + p014 + p015 + p018 + p019 +
p019 + p0110 + p0111 + p0112 + p0113 + p0114 + p0115 + p0116 + p0117 + p0118
+ p0119 + p0121 + p0122 + p0123 + p0124 + p0125 + p0126 + z01b.1 + z01b.2+
z01b.3 + z01b.4 + z01b.5 + z01b.6 + z01b.7 + z01b.8 + z01b.9 + z01b.10 + z01b.11
+ g10 + g11.1 + g11.2 + g11.3 + g11.4 + g11.5 + g01q.1 + g01q.2 + g01q.3 +
g01q.4 + g02.1 + g02.2 + g06.1 + g06.2 + g06.3+ g06.4, data=Epurchase.df,
control=temp1)
# FINAL TREE SELECTION
#Generate Dataframe with cp values for optimal subtrees
testcp<- data.frame(printcp(fit1))
#Look for average error from cross-validation and choose tree with minimum error
searchminxerror <- subset(testcp, testcp$xerror==min(testcp[,4]))
#Give cp value of min xerror
minxerrorcp<- searchminxerror [nrow(searchminxerror),1]
#Saturated Tree is trimmed with cp value corresponding to min xerror
FinalTree<-prune(fit1, cp=minxerrorcp)
#Summary data of Final Tree
Shortsummary<-nrow(printcp(FinalTree))
Detailedsummary<-(summary(FinalTree))
#EXTERNAL OUTPUT
#Excel File ("CP_Values.xls") with CP data is generated
write.matrix(testcp, file = "C:/Dokumente und
Einstellungen/Burkhard/Desktop/Diplomarbeit/Computational/R/Results/
CP_Values.xls", sep = ";")
#Postscript File ("Saturated Tree.ps" with Graphic Illustration of Saturated Tree is
generated
postscript(file=paste("C:/Dokumente und
Einstellungen/Burkhard/Desktop/Diplomarbeit/Computational/R/Results/Saturated
Tree.ps", sep=""), width=200, height=100)
plot(fit1, uniform=TRUE, branch=0, compress=FALSE, margin=0.1,
```

```
main="SaturatedTree")
text(fit1, all=TRUE, use.n=TRUE, cex=0.4)
dev.off()
# Postscript File with Graphic Illustration of Selected Optimal Tree
postscript(file=paste("C:/Dokumente und Einstellun-
gen/Burkhard/Desktop/Diplomarbeit/Computational/R/Results/FinalTree.ps", sep=""),
width=200, height=200)
plot(FinalTree, uniform=TRUE, branch=1, compress=FALSE, margin=0.1)
text(FinalTree, all=TRUE, use.n=TRUE, cex=0.4)
dev.off()
graphics.off()
# Calculation Time
End.Time<-Sys.time()
Calculation.Time<-difftime(End.Time, Start.Time)
Calculation.Time
```

6.5 Data Output for E-purchase Survival Tree

CP	nsplit	rel error	x error	x std
0.0199936	0	1.00000	1.00019	0.011555
0.0113987	1	0.98001	0.98039	0.011400
0.0089754	2	0.96861	0.96915	0.011519
0.0063544	3	0.95963	0.96027	0.011678
0.0050783	4	0.95328	0.95528	0.011737
0.0046397	5	0.94820	0.95036	0.011834
0.0039479	6	0.94356	0.94869	0.011858
0.0035478	8	0.93566	0.94247	0.011932
0.0035467	9	0.93212	0.94100	0.011926
0.0033221	10	0.92857	0.94102	0.011931
0.0025305	12	0.92193	0.92861	0.011879
0.0023094	13	0.91939	0.92725	0.011924
0.0022233	14	0.91709	0.92644	0.011942
0.0019823	15	0.91486	0.92561	0.011987
0.0018603	18	0.90892	0.92728	0.012031
0.0016113	20	0.90519	0.92511	0.012079
0.0015357	21	0.90358	0.92416	0.012084
0.0013623	23	0.90051	0.92232	0.012145
0.0013432	24	0.89915	0.92211	0.012150

6.6 Saturated Survival Tree for E-purchase Adoption

Saturated Tree

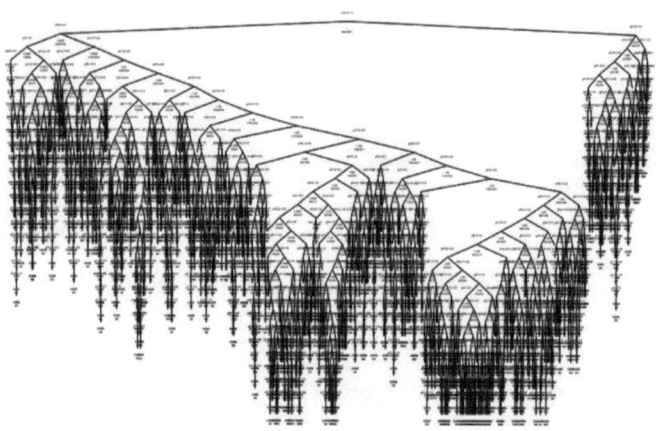

6.7 Original Survival Tree for E-purchase

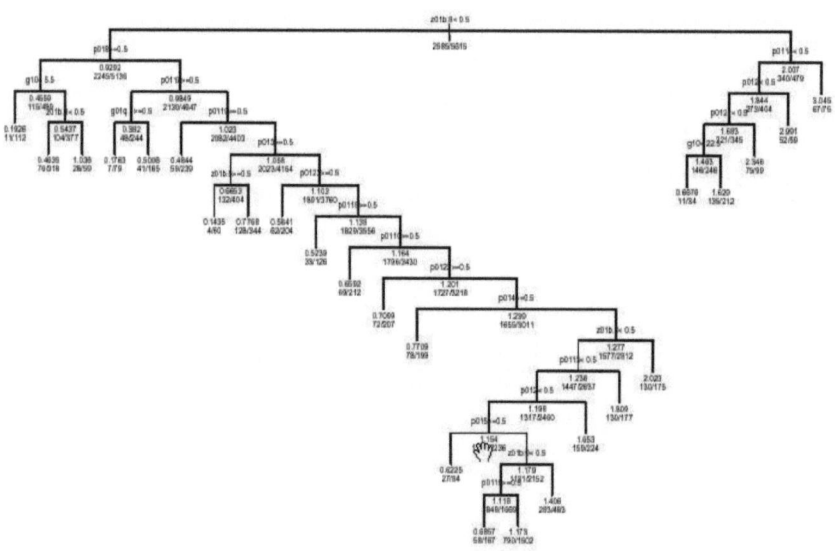

7 Indices

7.1 Index of Abbreviations

CART™   Classification and regression tree /software
ADT   Adoption- and diffusion theory

7.2 Index of Symbols

T	Total time
t_i	Duration of an episode for i-th member of the sample
τ	Tree node
τ_0	Saturated Tree
$r(\tau)$	Misclassification cost resulting from any subject within a terminal node with $\tau \in \tilde{T}$
$\tilde{\tau}$	Terminal node
$P(\tau)$	The probability of a subject to fall into the terminal node τ with $\tau \in \tilde{T}$
δ	Censoring indicator, $\delta=1$: uncensored, $\delta=0$: observed time censored
$R(T)$	Classification performance of entire tree
$R(\tau)$	Misclassification costs for node τ where $\tau \in \tilde{T}$
$R_\alpha(T)$	Cost complexity
$i(\tau)$	Impurity of a node
$h(t)$	Hazard rate
$h_0(t)$	Baseline hazard
X_i	Vector of covariates of the ith member of the sample
$x_1, ... x_p$	p single covariates
c	A constant
α	Complexity parameter
c	A constant
β	Vector of coefficients

7.3 Index of Synonyms

Hazard models:	Event history models, failure-time models, life-time models, survival models, transition rate models, response-time models, hazard rate models, duration model.
Event history data:	Survival data, censored event data, event history data, failure time data, duration data, transition data.
Event history analysis:	Survival analysis.
Duration:	Spell, waiting time, episode, duration of an episode.
Survivor function:	Survival function.
Co-variables:	Predictors, covariates, independent variables, explanatory variables.

7.4 Index of Tables

Table 2-1 Mean adoption time 28

Table 2-2 Log-rank 2x2 matrix 31

Table 2-3 Coefficients from semi-parametric Cox Model 37

Table 4-1 E-business adoption pattern for recorded dependent variables 70

Table 4-2: Summary of LeBlanc and Crowley (1992) proposal 72

Table 4-3: Data output for final tree 75

7.5 Index of Figures

Figure 2-1: Right- and left censored data in ADT 25

Figure 2-2: Kaplan-Meier estimate of the survivor functions
for adoption of 3 e-business solutions 30

Figure 2-3: Exponential distribution of survivor function 34

Figure 3-1: Classification and regression tree for e-purchase adoption 43

Figure 3-2: Three possible Kaplan-Meier curves for a homogeneous node........ 47

Figure 3-3: The L^1 Wasserstein distance between two Kaplan-Meier curves..... 48

Figure 4-1: 1-Survivor function for e-purchase adoption 70

Figure 4-2: Survival tree for e-purchase adoption 73

Figure 4-3: Log-Rank test for node 2 * node 3 76

Figure 4-4: Log-Rank test for Node 98244 * Node 98245 76

Figure 4-5: Log-Rank test for Node 13 * Rest 77

Figure 4-6: Log-rank test Node 20 * Rest 77

Figure 4-7: Scenarios with differing censoring structure 81

8 Bibliography

A.J. FEELDERS (1999), "Handling missing data in trees: Surrogate splits or statistical imputation?", *Proceedings of the third European conference on principles and practice of knowledge discovery in data bases (PKDD99), New York, Springer, pp. 329-334.*

AGRAWAL, R. AHUJA, M. CARTER, P. GANS, M. (1998), "Early and late adopters of IT innovations: Extensions of innovation diffusion theory", *Florida State University, Manuscript.*

ALLISON, P. (1984), "Event history analysis, regression for longitudinal event data", *Beverly Hills, Sage Publications, Newbury Parka.*

ALTMAN, D. ROYSTON, P. (2000), "What do we mean by validating a prognostic model?", *Statistics in Medicine, 19, pp. 453-473.*

ANDERSEN, P., KEIDING, N. (2001), "Event history analysis: selectivity", Manuscript, http://spitswww.uvt.nl/~vermunt/iesbs2001.pdf.

ANSELL ET. AL. (1993), "Survival determinants in patients with advanced ovarian cancer", *Gynecologic Oncology, 50, pp. 215-250.*

AUDRETSCH, D. (1991), "New firm survival and the technological regime", *The Review of Economics and Statistics, 60, 3, pp. 441-450.*

AUDRETSCH, D. MAHMOOD, T. (1995), "New firm survival: New results using a hazard function", *The Review of Economics and Statistics, 77, 1, pp. 97-103.*

BANDURA, A. (1997), "Self-efficacy: The exercise of control", *New York, Freeman.*

BAPTISTA, R. (2000), "Do innovations diffuse faster than within geographical clusters?", *International Journal of Industrial Organisation, Vol. 18, pp. 515-535.*

BLOSSFELD, H. HAMERLE, A. MAYER, K. (1986), "Ereignisanalyse: Statistische Theorie und Anwendung in den Wirtschafts- und Sozialwissenschaften", *Frankfurt a.M./New York, Campus.*

BERNDT, R. ALTOBELLI, C. (1991), "Warum Bildschirmtext in der Bundesrepublik scheiterte – Eine diffusionstheoretische Analyse einer verfehlten Marketing-Politik", *Zeitschrift für betriebswirtschaftliche Forschung, 43, pp. 955-969.*

BLOSSFELD, H. HAMERLE, A. MAYER, K. (1989), "Event history analysis", *Hillsdale, New Jersey, Erlbaum.*

BLOSSFELD, H. ROHWER, G. (2002), "Event history analysis", *2nd Edition, New Jersey, Lawrence Erlbaum Associates.*

BOSE, S. KOOPERBERG, C. STONE, C. (1995), "Polychotomous regression", *Manuscript,* http://stat.washington.edu/www/tech.reports/reports/1995/tr288.ps.

BREIMAN, L. (1996), "Bagging predictors", Machine Learning, 24, pp. 123-140.

BREIMAN, L. FRIEDMAN, J. OLSHEN, R. AND STONE, C. (1984), "Classification and regression trees", *Belmont, California, Wadsworth International Group.*

BRESLOW, N. (1972), "Contribution to the discussion of a paper by D. Cox", *Journal of the Royal Statistical Society, Series B, 34, pp. 216-217.*

BRIER, G. (1950), "Verification of weather forecasts expressed in terms of probability", *Monthly Weather Review, 78, pp. 1-3.*

BÜHLMANN, P. YU, B. (2002), "Analyzing bagging", *The Annals of Statistics, 30, pp. 927-961.*

BULT, J. (1993), "Target selection for direct marketing", *Ph.D Thesis, Netherlands, Rijksuniversiteit Groningen.*

BUTLER, J. GILPIN, E, GORDON, L. OLSHEN, R. (1991), "Tree structured survival analysis", *Second Technical Report, Stanford University.*

BYAR D. (1988), "Identification of prognostic factors", *Cancer Clinical Trials: Methods and Practice, Oxford University Press, Oxford, pp. 423-443.*

CATEORA, P. (1993), "International Marketing", *The Irwin Series in Marketing, 8th Edition, New York.*

CHATTERJEE, R. ELIASHBERG, J. (1989), " The innovation diffusion process in a heterogeneous population: A micro-modelling approach", *Working paper, Marketing Department, Krannert Graduate School of Management, Purdue University.*

CIAMPI, A. THIFFAULT, J. NAKACHE, J. ASSELAIN, B. (1986), "Stratification by stepwise regression, correspondence analysis and recursive partitioning", *Computational Statistics and Data Analysis, 4, pp. 185-204.*

CLARK, L., PREGIBON, D. (1992), "Tree-based models", *Chambers, J. Hastie, T. (Eds.), Statistical models in S, Pacific Grove.*

COX, D. (1972), "Regression model and life tables", *Journal of Royal Statistical Society, Series B 34, pp. 187-202.*

COX, D. OAKES, D. (1984), "Analysis of survival data", *London, New York, Chapman and Hall.*

COX, D. (1995), "Some Remarks on the analysis of survival data", *Manuscript.*

CROWLEY, J. LEBLANC, M. GENTLEMAN, R. SALOMON, S. (1995), "Exploratory methods in survival analysis. *Analysis of censored data", IMS lecture notes-monograph series 27, H.L. Koul and J. Deshpande, Eds, pp. 55-77, Institute of Mathematical Statistics, California, Hayward.*

CURRAN ET. AL. (1993), "Recursive partitioning analysis of prognostic factors in three radiation therapy oncology group malignant glioma trails", *Journal of the National Cancer Institute, 85, pp. 704-710.*

DANNEGGER, F. (2000), "Tree stability diagnostics and some remedies for instability", *Statistics in Medicine, 19, pp. 475-491.*

DAVIS, R. ANDERSON, J. (1989), "Exponential Survival Trees", *Statistics in Medicine, 8, pp. 947-962.*

DE ROSE, R. PALLARA, A. (1997), "Survival trees: an alternative non-parametric multivariate technique for life history analysis". *European Journal of Population, 13, pp. 223-241.*

EFRON, B. TIBSHIRANI, R. (1993), "An introduction to the bootstrap", *London, Chapman & Hall.*

ENQUIST, C. (1997), "Systems of innovation approaches - their emergence and characteristics", *Systems of Innovation: Technologies, Institutions and Organizations, Chapter one, London and Washington, pp. 1-35.*

EVANS, D. EVERIS, L. BETTS, G. (2004), "Use of survival analysis and classification and regression trees to model the growth/no growth boundary of spoilage yeasts as affected by alcohol, pH, sucrose, sorbate and temperature", *International Journal of Food Microbiology, 92, pp. 55-67.*

Faraggi, D. LeBlanc, M. Crowley, J. (2001), "Understanding neural networks using regression trees: An application to multiple myeloma survival data", *Statistics in Medicine, 20, pp.2965-2976.*

FARAGGI, D. SIMON, R. (1994) "A neural network model for survival data", *Statistics in Medicine, 14, pp. 73-82.*

FEDER, G. O'MARA (1982), "On information and innovation diffusion: A Bayesian approach", *American Journal of Agricultural Economics, 64, pp. 145-147.*

FEELDERS, A. CHANG, S. MCLACHLAN, G. (1998),"Mining in the presence of selectivity bias and its application to reject inference", *Proceedings of the Fourth International Conference on Knowledge Discovery and Data Mining, Menlo Park, California, AAAI Press, pp. 199-203.*

FISHBEIN, M. AJZEN, I. (1975), "Belief attitude, intention, and behaviour: An introduction to theory and research", *Addison-Wesley, Reading.*

FISHBEIN, M. AJZEN, I. (1980), "Understanding attitudes and predicting social behaviour", New Jersey, *Prentice-Hall.*

FOX, J. (2003), "Cox proportional-hazards regression for survival data", *Manuscript,*
http://socserv.mcmaster.ca/jfox/Books/Companion/appendix-cox-regression.pdf.

FRIEDMAN, J. (1991), "Multivariate regression splines (with discussion)", *The Annals of Statistics, 19, pp. 100-14.*

FRIEDMAN, M. (1985), "Technical report 12", *Department of Statistics, Stanford University.*

GEROSKI, P. (2000), "Models of technology diffusion", *Research Policy.*

GORDON, L. OLSHEN, R. (1984), "Almost surely consistent non-parametric regression from recursive partitioning schemes", *Multivariate Analysis, 15, pp. 147-163.*

GORDON, L. OLSHEN, R. (1985), "Tree-structure survival analysis", *Cancer Treatment Reports, 69, pp. 1065-1069.*

GOURLAT, A. PENTECOST, E. (2000), "The determinants of technology diffusion: Evidence from the UK financial sector", *Economic Research Paper No. 00/9, Department of Economics, Loughborough University.*

GRAF, E. SCHMOOR, C. SAUERBREI W. SCHUMACHER, M. (1999), "Assessment and comparison of prognostic classification schemes for survival data", *Statistics in Medicine, 18, pp. 2529-2545.*

GRILICHES, Z. (1957), "Hyprid corn: an exploration in the economics of technological change", *Econometricam, Vol. 25, pp. 501-522.*

HANNAN, T. MCDOWELL, J. (1984), "The determinants of technology adoption: the case of the banking firm", *Rand journal of Economics, 15, pp. 328-335.*

HANNAN, T. MCDOWELL, J. (1987), "Rival precedence and the dynamics of technology adoption: an empirical analysis", *Economica, Vol. 54, pp. 155-171.*

HAUGHTON, D. OULABI, S. (1993), "Direct marketing modelling with CART and CHAID", *Journal of Direct Marketing, 7, pp. 16-26.*

HECKMAN, J. SINGER, B. (1974), "Econometric Duration Analysis", *Journal of Econometrics, 24, pp. 63-132.*

HECKMAN, J. SINGER, B. (1984), "The identifiability of the proportional hazard model", *Review of Economic Studies, 51, pp. 231-241.*

HECKMANN, J. BORJAS, G. (1980), "Does unemployment cause future unemployment? Definitions, questions and answers from a continuous time model of heterogeneity and state dependence", *Econometrica, 47, pp. 247-283.*

HENDERSON, R. (1995), "Problems and prediction in survival data analysis", *Statistics in Medicine, 14, pp. 161-184.*

HIEBERT, L. (1974), "Risk learning and the adoption of fertilizer responsive seed varieties", *American Journal of Agricultural Economics, 56, pp. 764-768.*

HOLLANDER, M. PROSCHAN, F. (1979), "Testing to determine the underlying distribution using randomly censored data", *Biometrics, 35, pp. 393-401.*

HONJO, Y. (2000), "Business failure of new firms: An empirical analysis using multiplicative hazard models", *International journal of industrial organization, 18, 4, pp. 557-574.*

HORNSTEINER, U. HAMERLE, A. MICHELS, P. (1997), "Parametric versus nonparametric treatment of unobserved heterogeneity in multivariate failure times", *Manuscript,*
http://www.stat.uni-muenchen.de/sfb386/papers/dsp/paper80.ps.

HOSMER, D. LEMESHOW, S. (1999), "Applied survival analysis", *New York, John Wiley and Sons.*

HOTHORN, T. LAUSEN, B. BENNER, A. RADESPIEL-TROEGER, M. (2004), "Bagging survival trees", *Statistics in Medicine, 23, pp. 77–91,* www.mathpreprints.com/math/Preprint/blausen/20020518/2/bstpp.pdf.

INGHAM, H. THOMPSON, S. (1993), "The adoption of new technology in financial services: The case of building societies", *Economics of Innovation and New Technology, 2, pp. 263-274.*

INTRATOR, O. (1991), "Exploratory trees for semi-markov processes", *Proceedings of the 23rd Symposium on the Interface of Computing Science and Statistics, pp. 352-355.*

INTRATOR, O. KOOPERBERG, C. (1995), "Trees and splines in survival analysis", *Statistical Methods in Medical Research, 4, 3, pp. 237-262.*

KALBFLEISCH, J. PRENTICE, R. (1980), "The statistical analysis of failure time data", *New York, John Wiley and Sons.*

KAPLAN, E. MEIER, P. (1958), "Nonparametric estimation from incomplete estimations", *Journal of the American Statistical Association, 53, pp. 457-481*

KAUFMAN, R. WANG, B. (2001), "The success and failure of dotcoms: A multi-method survival analysis", *Manuscript,* http://misrc.umn.edu/workingpapers/fullPapers/2001/0109_092501.pdf.

KELES, S. SEGAL, M. (2002), "Residual-Based Tree structured survival analysis", *Statistics in Medicine, 21, pp. 313-326.*

KLEINBAUM, D. (1995), "Survival analysis - a self learning text", *New York, Springer.*

KÖLLINGER, P. SCHADE, C. (2004), " Adoption of e-business: patterns and consequences of network externalities", *Manuscript,*
http://www.case.hu-berlin.de/index_html/Publikationen/papers/papersKatalog/05_pk_cs.pdf.

KOOPERBERG, C. STONE, C. (1992), "Logspline density estimation for censored data", *Journal of Computational and graphical Statistics, 1, pp. 301-328.*

KOOPERBERG, C. STONE, C. TRUONG, Y. (1995), "Hazard regression", *Journal of the American Statistical Association, 90, pp. 78-94.*

KORN, E. SIMON, R. (2000), "Measures of explained variation for survival data", *Statistics in Medicine, 9, pp. 487-503.*

LANCASTER, L. (1990), "The econometric analysis of transition data", Cambridge, *Cambridge University Press.*

LATTIN, J. ROBERTS, J. (1989), "Modelling the role of risk-adjusted utility in the diffusion of innovations", *Working Papers, 1019, Graduate School of Business, Stanford University.*

LAUSEN, B. SAUERBREI, W. SCHUMACHER M. (1994), "Classification and regression trees used for exploration of prognostic factors measured on different scales", *Computational statistics, P. Dirschedl and R. Ostermann (Eds.), Physica-Verlag Heidelberg, pp- 483-496.*

LEBLANC M. (2001), "Tree-based methods for prognostic stratification", *Statistics in Oncology, Crowley, J. (ed.) Marcel Dekker: New York, Basel, 2001, pp. 457-472.*

LEBLANC, M. CROWLEY, J. (1992), "Relative risk trees for censored survival data", *Biometrics, 48, pp. 457-467.*

LEBLANC, M. CROWLEY, J. (1993), "Survival trees by goodness of split", *Journal of the American Statistical Association, 88, pp. 457-467.*

LEBLANC, M. CROWLEY, J. (1995), "A review of tree based prognostic models", *New Advances in the Design and Analysis of Clinical Trials Data, Kluweer Academic Publishers.*

LEBLANC, M. CROWLEY, J. GENTLEMAN, R. SALOMON, S. (1996), "Some exploratory methods in survival analysis", *Manuscript.*

LITFIN, T. (2000), "Adoptionsfaktoren: Empirische Analyse am Beispiel eines innovativen Telekommunikationsdienstes: Mit einem Geleitwort von Sönke Albers", *Wiesbaden, Manuscript.*

LEE, E. (1992), "Statistical methods for survival analysis", *2nd, Wiley and Sons.*
LILIEN, G. KOTLER, P. MOORTHY, S. (1992), "Marketing Models", *USA, Prentice-Hall.*

MAHAJAN, V. MULLER, E. (1979), "Innovation diffusion and new product growth models in marketing", *Journal of marketing, 43, pp. 55-68.*

MAHAJAN, V. PETERSON, R. (1985), "Models for innovation diffusion", *Quantitative Applications in the Social Sciences, Newbury Park, London, New Delhi, Sage.*

MAHAJAN, V. MULLER, E. BASS, F. (1990), "New product diffusion models in marketing: A review and directions for research", *Journal of Marketing, 54, 1, pp. 1-26.*

MANSFIELD, E. (1961), "Technical Change and the rate of imitation", *Econometrica, 29, pp. 741-766.*

MATA, J. PORTIGAL, P. GUIMARÄES, P. "The survival of new plants start-up conditions and post entry evolution", *International Journal of Industrial Organisation, 13, pp. 459-481.*

MAYER, K. TUMA, N. (1990), "Event history analysis in life course research", *USA, The University of Wisconsin Press.*

MENARD, S. (1995), "Applied logistic regression analysis", California, *Thousand Oaks.*

MILLER,R. (1981), "Survival analysis", *New York. Wiley and Sons.*

MORGAN, J. SONQUIST, J. (1963), "Problems in the analysis of censored data and a proposal", *Journal of the American Statistical Association, 58, pp. 415-435.*

NELSON, W. (1969), "On estimating the distribution of a random vector when only the coordinate is observable", *Technometrics, 12, pp.923-924.*

NOBEL, A. (2000), "Analysis of a complexity based pruning scheme for classification trees", *Manuscript.*

OREN, S. SCHWARTZ, R. (1988), "Diffusion of new products in risk-sensitive markets", *Journal of Forecasting, 7, pp. 273-28.*

PETER, P. OLSON, C. (2002), "Consumer behaviour and marketing strategy", *6th Edition, New York, McGraw-Hill/Irwin.*

QUIRMBACH, H. (1986), "The diffusion of new technology and the market for and innovation", *Rand Journal of Economics, 17, pp. 33-47.*

RADESPIEL-TRÖEGER, M. RABENSTEIN, H. LAUSEN, B. (2003a), "Comparing splitting algorithms in survival trees using brier score for censored data", *Manuscript.*

RADESPIEL-TRÖEGER, M. RABENSTEIN, T. SCHNEIDER, H. LAUSEN, B. (2003b), "Comparison of tree-based methods for prognostic stratification of survival data", *Elsevier Science.*

REINGANUM, J. (1982) "A dynamic game of R&D: Patent protection and competitive behaviour", *Econometrica, 50, pp. 671-688.*

ROGERS, E. (1995), "Diffusion of innovations", *4th Edition, New York, The Free Press.*

SALONER, G. SHEPARD, A. (1995), "Adoption of technologies with network effects: an empirical examination of the adoption of automated teller machines", *Rand Journal of Economics, Vol. 26, pp. 479-501.*

SARKAR, J. (1998). "Technological diffusion: alternative theories and historical evidence", Journal of Economic Surveys, 12, pp. 131-176.

SAUERBREI, W. (1998), "Boot-strapping in survival analysis", *Encyclopedia of Biostatistics, Armitage P, Colton T (Eds), New York, Wiley and Sons, pp. 433-436*

SCHAPIRE, R. (1999), "Theoretical views of boosting", *Computational Learning Theory: Fourth European Conference, EuroCOLT99, pp. 1-10.*

SCHAPIRE, R. (2002), " The boosting approach to machine learning: An overview", *MSRI Workshop on Nonlinear Estimation and Classification.*

SCHLITTGEN, R. (1999), "Regression trees for survival data - an approach to select discontinuous split points by rank statistics", *Biometrical Journal, 41, 8, pp. 943 – 954.*

SCHUMACHER, M. HOLLÄNDER, N. SCHWARZER, G. SAUERBREI, W. (2001), "Prognostic factor studies", *Statistics in Oncology, Crowley J. (ed.) Marcel Dekker: New York, Basel, pp. 321-378.*

SEGAL, M, BLOCH, D. (1989), "A comparison of estimated proportional hazard models and regression trees", *Statistics in Medicine, 8, pp. 539-550.*

SEGAL, M. (1988), "Regression trees for censored data", *Biometrics, 44, pp. 35-47.*

SETNES, M. KAYMAK, U. (2001), "Fuzzy modelling of client preference from large datasets: an application to target selection in direct marketing", *Manuscript.*

SINHA, R. CHANDRASHEKARAN M. (1992), "A split hazard model for analysing the diffusion of innovations", *Journal of Marketing, 29, pp. 116-127.*

STONEMAN, P. (1981), "Intra-firm diffusion, Bayesian learning and probability", *Economic Journal, 91, pp. 375-388.*

STONEMAN, P. (1995), "Handbook of the economics of innovation and technological change", *Oxford, Cambridge, Blackwell Publishers.*

SU, X. FAN, J. (2001), "Multivariate survival trees by goodness of split", *Technical Report 367, Department of Statistics, University of California, Davis.*

SU, X. FAN, J. (2004), "Multivariate survival trees: A maximum likelihood approach based on frailty models", *Biometrics, 60, pp. 93-99.*

TAYLOR, J. COBB, K. (2004), "Statistics 262: Intermediate Biostatistics", *Lecture notes,*
http://www-stat.stanford.edu/~jtaylo/courses/stats262/spring.2004/notes/week5.pdf.

THERNEAU, T. ATKINSON, E. (1997), "An introduction to recursive partitioning using the rpart routine", *Technical Report 61, Section of Biostatistics, Rochester, Mayo Clinic.*

THERNEAU, T. GRABSCH, P. FLEMING, T. (1990), "Martingale based residuals for survival models", *Biometrika 77, pp. 147-160.*

TIBSHIRANI, R. KNIGHT, K. (1995), "Model search and inference by bootstrap burning", *Technical Report, Department of Statistics and Preventive Medicine and Biostatistics, University of Toronto.*

ULM, K. PASHOVA, V. (2000), "Two survival tree methods for Myocardial infarction patients", *Manuscript,*
http://www.stat.uni-muenchen.de/sfb386/papers/dsp/paper195.ps

VAN BRUGGEN, G. SMIDTS, A. WIERENGA, B. (2000), "The powerful triangle of marketing data, managerial judgement and marketing management support systems", *Research in Management.*

VENABLES, W. RIPLEY B. (1994), "Modern applied statistics with s-plus", *2nd Edition, New York, Springer, New York.*

WIRIGHT, M. CHARRIET, D. (1995), "New product diffusion models in marketing: An assessment of two approaches", *Marketing Bulletin, 6, pp. 32-41.*

ZAHAVI, J. LEVIN, N. (1997), "Applying neural computing to target marketing", *Journal of Direct Marketing, 11, 1, pp. 5-22.*

ZHANG, H. (1995), "Splitting criteria in survival trees", *10th Workshop on Statistical Modelling, Lecture Nodes in Statistics Series, New York, Springer, pp. 305-314.*

ZHANG, H. (1998), "Classification trees for multiple binary responses", *Journal of the American Statistical Association, 93, pp. 180-193.*

ZHANG, H. SINGER, B. (1999), "Recursive partitioning in the health sciences", *New York, Springer.*

ZHANG, H. BRACKEN, M. (1995), "Tree-based risk factor analysis of preterm delivery and small-for-gestational-age birth", *American Journal of Epidemiology, 141, pp. 70-78.*

ZHANG, H. BRACKEN, M. (1996), "Tree-based, two-stage risk factor analysis for spontaneous abortion", *American Journal of Epidemiology, 144, pp. 989–996*

ZHANG, H.(1999), "Recursive partitioning and tree-based methods", M*anuscript,*
http://www.case.hu-berlin.de/index html/Publikationen/papers/papersKatalog/30 hz